Miniature Sorption Coolers

Theory and Applications

Miniature Sorption Coolers

Theory and Applications

L. Piccirillo
G. Coppi
A. May

CRC Press
Taylor & Francis Group
Boca Raton London New York

CRC Press is an imprint of the
Taylor & Francis Group, an **informa** business

CRC Press
Taylor & Francis Group
6000 Broken Sound Parkway NW, Suite 300
Boca Raton, FL 33487-2742

Printed on acid-free paper
Version Date: 20180206

International Standard Book Number-13: 978-1-4822-6041-0 (Hardback)

Library of Congress Cataloging-in-Publication Data

Names: Piccirillo, Lucio, author. | Coppi, Gabriele, author. | May, Andrew (PhD student), author.
Title: Miniature sorption coolers : theory and applications / Lucio Piccirillo, Gabriele Coppi, Andrew May.
Description: Boca Raton, FL : CRC Press, Taylor & Francis Group, [2018] | Includes bibliographical references and index.
Identifiers: LCCN 2017043054| ISBN 9781482260410 (hardback ; alk. paper) | ISBN 9781351188678 (e-book general) | ISBN 9781351188654 (e-book epub) | ISBN 9781351188661 (e-book pdf) | ISBN 9781351188647 (ebook mobipocket)
Subjects: LCSH: Low temperature engineering. | Refrigeration and refrigerating machinery.
| Image analysis--Mathematical models.
Classification: LCC TP480 .P53 2018 | DDC 621.5/6--dc23
LC record available at https://lccn.loc.gov/2017043054

Visit the Taylor & Francis Web site at
http://www.taylorandfrancis.com

and the CRC Press Web site at

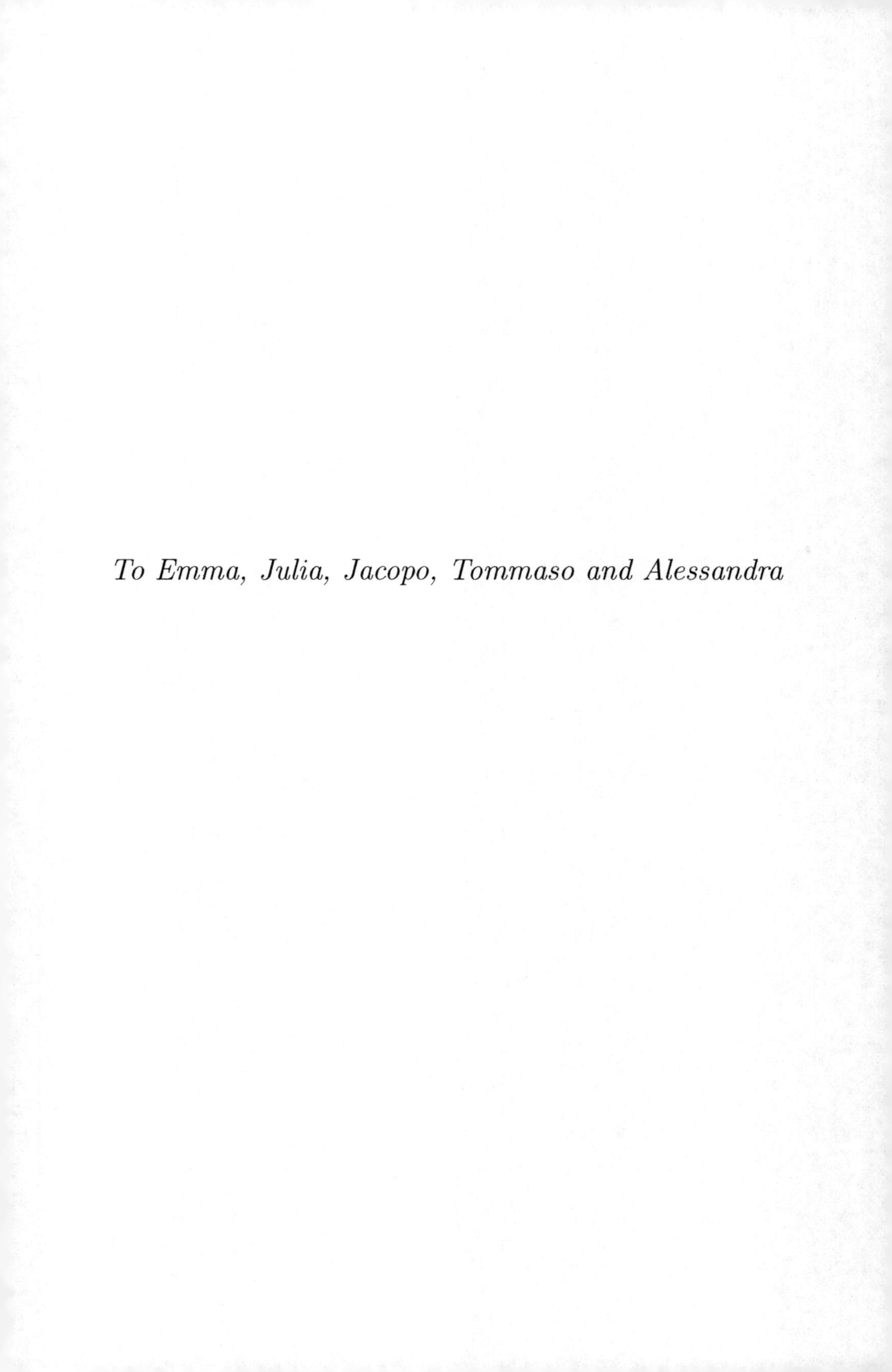

To Emma, Julia, Jacopo, Tommaso and Alessandra

Contents

List of Figures

List of Tables

Preface

In the course of our research in the field of mm-wave experimental astrophysics, we have faced constant challenges in cooling to sub-K temperatures the detectors (usually bolometers) needed to reach the sensitivities required by the observations. Our research necessitated the deployment of special mm-wave telescopes to remote, high-altitude sites, from the Antarctic plateau to the Atacama high-altitude desert in the Chilean Andes. These telescopes are already exceptionally sophisticated machines and hence the additional complexity of cryogenic detectors makes them particularly challenging from an engineering design and operational perspective. Considering that these experiments are designed to run for years in virtually inaccessible sites at high altitudes, it is clear that our collective efforts have been always oriented towards designing systems which, whilst being compact, lightweight and providing high cooling power, require limited external machinery (such as pumps and compressors) and may be fully computer controlled.

A significant simplification occurred when mechanical cryocoolers first entered the market capable of providing cooling powers of $\sim$ 1 W at 4 K and several tens of W at 50 K. This small "revolution" freed us from using liquid helium as a cold reservoir, with all of the logistical issues associated with its use such as transportation to the site, periodic liquid helium transfer into the detector, etc. It was immediately evident that if, with suitable vibration mitigation, we could built cryostats around 4 K mechanical coolers and then equip these with the further sub-K cooling stages needed for operating the detector arrays, then we would simplify enormously the running of our experiments.

We have now built several small closed-cycle sub-K refrigerators for various operating temperatures, from 1 K single-stage ^{4}He system to 50 mK continuous miniature dilution refrigerators. In the course of the painful road leading to the construction of working systems, we have collected a great deal of experience through many failures and a smaller number of successes.

Given the wider range of applications that these types of coolers have now found, we decided to write this book in order to collect, in a single place, all of the necessary information, both theoretical and practical, to allow any good experimental physicist or engineer to successfully design and build a custom sub-K closed-cycle refrigerator. We do not claim that our proposed systems are the best and most optimized refrigerators that can possibly be designed and built. However, we do claim that the reader of this book *can* build his/her working system without relying on the expensive and time consuming commercial route — especially when whatever is available on the market is not exactly what is needed for their application.

If at least one successful system is built based on the recipes and principles discussed in this book, then all of the many hours of discussions, writing, correcting, fighting with the editor because we are late, and many more, will have been justified.

Lucio Piccirillo, Gabriele Coppi and Andrew J. May
Manchester, January 2018

Authors

Lucio Piccirillo is Professor of Radio Astronomy Technology at the Jodrell Bank Centre for Astrophysics at the University of Manchester, UK. He is an expert in experimental cosmology, cryogenics, RF devices and novel astrophysical systems. He has more than 130 publications.

Gabriele Coppi is a PhD student in Astronomy and Astrophysics at the University of Manchester. He did his MSc at University La Sapienza in Rome, developing the focal plane of the LSPE/SWIPE experiment. He is now working on cryogenics instrumentation and design for some ground-based CMB experiments.

Andrew J. May is a PhD student in Astrophysics at the University of Manchester, where he also completed his undergraduate and master's degrees. During his undergraduate work, he also worked at UCL's Mullard Space Science Laboratory and the STFC Daresbury Laboratory on the development of cryogenic systems for the Large Hadron Collider upgrade. He then came to JBCA to work on sub-Kelvin systems under Lucio Piccirillo.

I

Background and Theory

CHAPTER 1

Background

THE FIELD of cryogenics, by convention [73] defined as the branch of science and technology of temperatures below 120 Kelvin[1], is one that can claim a great history of success. A considerable number of technological innovations have allowed the production of increasingly lower temperatures and these have found an incredibly broad range of applications across many fields of human endeavour.

One of the first applications of refrigeration technology was in that most fundamental of human activities: eating food. Low temperatures are needed to preserve food from spoiling because the low temperature somehow "freezes" all biological activities including the bacteria responsible for food decay. The more technologically advanced the human race has become, the more advanced have become the activities supported by low-temperature technologies. Think, for example, of the liquid oxygen engines used to fuel space rockets. In medicine, liquid helium temperatures are today routinely used in MRI scanning machines in most hospitals around the world to allow the production of intense magnetic fields for magnetic imaging that are a vital technique of modern medicine.

To find natural low temperatures, we need to look into deep space. The Universe itself is a blackbody cavity at a temperature of 2.7 K, relics of a time when the Universe was in equilibrium with radiation. Due to the cosmic expansion, this thermal radiation has cooled to the current thermodynamic temperature. The largest ever refrigerator is the Universe itself which is using its expansion to cool the thermal radiation contained in it. It is somehow remarkable that one of the most active communities using ultra-low temperatures are the astrophysicists looking to study the extremely high temperature conditions believed to exist at the time of the Big Bang. Temperatures below 1 K are used to cool bolometric detectors in order to reach the sensitivities required by modern experiments, today reaching the regime where the noise

[1]The upper limit of 120 K is chosen so as to include the boiling point under standard pressure of the main atmospheric gases, as well as methane (the main component of natural gas).

is limited only by quantum mechanics. The production of low temperature is fundamental in allowing scientists to achieve the best possible radiation detectors for studying our cosmic origins.

As physicists, we note that the laws of thermodynamics are one of the most powerful "tools" in our toolbox for understanding systems on a range of scales from Newtonian to cosmological. In writing this book, we have assumed of the reader a reasonable background in classical thermodynamics. However, detailed explanation of the key concepts is given where appropriate.

We need to be very careful when we apply thermodynamics to any system. A careful analysis of the forces in play needs to be done. Think for example when electric or magnetic fields are present which can give an additional degree of freedom and the proper thermodynamic quantities need to be calculated accordingly. We cannot resist to mention in here the thermodynamics of the Universe that was also mentioned above. Lord Kelvin, one of the fathers of thermodynamics, realized that the second principle of thermodynamics will imply that, waiting long enough, the entropy of the Universe will reach a maximum and there will be thermal equilibrium everywhere. This means that all the phenomena implying time evolution will at a certain point stop and stabilize on an equilibrium condition with disappearance of potential energy. Life is one of these phenomena.

Freeman Dyson confuted this view in a letter to *The New York Review of Books* discussing a book by James Gleick titled *The Information: a History, a Theory, a Flood* [29]. Freeman Dyson confuted Kelvin's heath death by stating that the heath death was considering only what he called **the cooking rule**. The cooking rule simply says that a steak gets warmer if it is on a hot grill. More generally, an object gets warmer when it gains energy or gets colder if it loses energy. Steaks have been cooked for thousands of years and not a single time a steak became colder when put on a hot grill. Now, if the cooking rule is correct, then Kelvin's heath death is inevitable. However, Dyson states that if we consider systems that are of astronomical size, then the cooking rule does not apply anymore because we have to consider another force which can be dominant: gravity. Consider for example the Sun: the Sun is losing constantly energy by radiation and it gets hotter and not cooler. Gravitation has a different relationship between energy and temperature: gravitational systems have negative heat capacity.[2] Therefore temperature differences, for astronomical systems, tend to increase rather than decrease or, in other words,

[2]The Virial theorem for a gravitationally bound system states that $\langle U \rangle = -\langle K \rangle$ where $\langle U \rangle$ is the average total energy and $\langle K \rangle$ is the average kinetic energy. Since the kinetic energy is always positive, it follows that gravitationally bound systems have always negative total energy. One remarkable consequence is that if we increase the total energy by an amount such that it is still negative, the kinetic energy must decrease. Inside a star, the temperature is high enough that particles have only translational degrees of freedom and the equipartition theorem tells us that each particle has an average kinetic energy equal to $\frac{3}{2}NkT$. This means that the heat capacity $C = \frac{\partial U}{\partial T} = -\frac{3}{2}Nk$. Beckenstein [7] and Hawking [33], [34] have also shown that black-holes have negative specific heat. See D. and R.M Lynden-Bell [53] for an extended discussion of the paradox.

if energy flows from a hot body to cold body the hot body will get hotter and the cold body will get colder. So there is not a state of equilibrium on an astronomical scale and so there is no heath death.

This is not the end of the story, though. Today we know that the Universe is not only expanding, but the rate of expansion is increasing due to a mysterious *dark energy*. So eventually everything will be torn apart... maybe. The story is still being written.

But let's go back to our systems where gravity is practically negligible. A wide range of techniques for refrigeration have been developed over the last century or so; clearly the choice of technique is constrained principally by the temperature required, as well as by other considerations such as the required heat lift, cooling time, available power, allowable level of vibration, and so forth.

Miniature closed-cycle sorption coolers, operating in the range 100 mK $\leq$ T $\leq$ 1 K, have been developed for cooling superconducting detectors and cold optics for various astrophysical applications [88, 18, 14, 22, 9, 20, 85, 60, 56, 16] and more recently have also found applications in other areas such as electron paramagnetic resonance instruments [62]. This book provides a detailed review of the theory, design and application of miniature sorption coolers.

The coolers which are the subject of this book operate in a closed cycle and take advantage of evaporative cooling of either ^{3}He or ^{4}He (depending on the required temperature) or dilution cooling of ^{3}He in ^{4}He. In order to discuss the design, analysis and operation of such coolers, it is first necessary to review the relevant thermodynamics, properties of matter at low temperatures and effects arising from the quantum mechanical behaviour of helium.

1.1 REVIEW OF RELEVANT THERMODYNAMICS

Thermodynamics became a serious branch of physics when scientists and engineers needed to understand the behaviour of steam engines. Steam engines are devices where heat and work play an important role and a lot of efforts were dedicated to set up the mathematical basis to describe them. Fundamental principles were established and the concept of entropy was immediately introduced. It then became clear, thanks mostly to the work of Boltzmann, that there was a close relationship between statistics and entropy and so the kinetic theory of matter was born. It is interesting to note that the vast majority of textbooks on thermodynamics and statistical physics start always by studying first the so-called "classical thermodynamics" and then prove most of the results by statistical physics. It is also possible to start with statistical physics and kinetic theory and then obtain the "classical thermodynamics" results. In this book we do not intend to give a comprehensive treatment of thermodynamics and statistical physics: many excellent books are dedicated to this topic. We will rather adopt a pragmatic approach and use results and techniques from both branches of mathematical physics on a basis of pure convenience. When statistical physics gives a better picture or solves a problem quicker, we will use it unless classical thermodynamics does a better job. Quite often, it does.

1.1.1 Differentials in thermodynamics

The systems used in the practice of sorption coolers have, as basic variables, the quantities P (pressure), V (volume) and T (temperature). Since we will be dealing mostly with equilibrium conditions, the fluids considered will be subject to an equation of state, i.e., a relationship between P, V and T. For example, in an ideal gas, we have that $PV = nRT$, where n are the number of moles and R is the gas constant ($R = 8.31446 J \cdot mol^{-1} \cdot K^{-1}$). In the absence of other external variables like, for example, a magnetic field, then any thermodynamic quantity can be expressed as a function of two variables chosen from P, V and T. The process of cooling can be regarded as a process by which the system is subject to a transformation where we properly vary one or more of the variables along some path in the P, V or P, T or V, T plane (sometimes called "state plane"). We will encounter many functions like heat Q, work W, internal energy U, entropy S, enthalpy H, etc. and we will study how such functions change during the transformations. A *state function* is a quantity that describe the equilibrium state of a system irrespective of the path followed to reach such equilibrium.[3] Entropy, internal energy and enthalpy (defined later in the book) are state functions, while heat and work are not. In the process of calculating quantities at the end of some thermodynamic transformation, we need to express the differentials of such quantities and then integrate along the path followed by the transformation. The power of state functions resides in the fact that the value of the integral depends only on the initial and final state. We will write the differential of a generic state function $f = f(x, y)$ in the state plane (x, y) with the symbol "*d*" to indicate that such a differential is "exact" or, in other terms:

$$df = \left(\frac{\partial f}{\partial x}\right)_y dx + \left(\frac{\partial f}{\partial y}\right)_x dy \tag{1.1}$$

Eq. 1.1 tells us immediately that f can be integrated along any path in the (x, y) state plane and that the result depends only on the initial and final states. Functions for which the differential *cannot* be written as Eq. 1.1 will be indicated with the symbol δf meaning that in order to calculate the final value during a transformation, we need to specify the path in the (x, y) state plane. Euler's test gives us a way to verify if a certain differential is exact or not. A function $f(x, y)$ of the two variables (x, y) is a state function if:

$$\left(\frac{\partial^2 f}{\partial x \partial y}\right) = \left(\frac{\partial^2 f}{\partial y \partial x}\right) \tag{1.2}$$

[3]There is a clear analogy with conservative forces and potential energy in classical mechanics.

It follows that, if we have functions that depend only on state functions, then they are also state functions and their differentials are exact.

1.1.2 The four laws of thermodynamics

Einstein [26] wrote that "[classical thermodynamics] is the only physical theory of universal content which I am convinced will never be overthrown, within the framework of applicability of its basic concepts."

Classical thermodynamics has a general validity in the physical science because of its simple laws and basic assumptions. Thermodynamics deals with the mutual transformations between heat and work. Having realized that heat is the result of the highly disordered motion of atoms or molecules, we would expect that thermodynamics should be derived by equations of motions of all the particles. Unfortunately, any system that has any macroscopic interaction — like the systems studied in this book, for example — is composed by a huge amount of atoms or molecules and solving the equations of motion is practically impossible. We therefore must forget the detailed knowledge of the state of a system and turn to average values. This way, to describe thermodynamics via statistical quantities is referred to as *statistical mechanics* first developed by Boltzmann, Maxwell and Gibbs two centuries ago. In the simple (pure) thermodynamics used in this book, we rarely go into the kinetic mechanisms but rather assume that the fundamental laws are given as postulates based on experimental evidence. In this way, thermodynamic results are very general and as accurate as the assumed postulates.

1.1.2.1 The first law of thermodynamics

The first law of thermodynamics expresses the conservation of energy for heat and thermodynamic processes. In terms of heat Q, internal energy U and work W the first law is written as:

$$\Delta U = Q + W \tag{1.3}$$

i.e., the variation of internal energy in a system is equal to the heat added *to* the system plus the work done *on* the system.[4] The internal energy U is the total energy contained in the system without considering its whole kinetic energy (for example, due to the whole motion of the system) and potential energy (for example, due to an external force). The first law 1.3 tells us that the internal energy can be changed by transferring heat or doing work. For those systems where matter can be exchanged, then we need to consider the contribution of the matter transferred to/from the system. In analogy with the case of the heat death of the Universe discussed above, we need to be

[4]Note that some chemistry books use a different convention for the sign of the work done by the system: $\Delta U = Q - W$ where now the "minus" sign is for the work done "by" the system. The two formulation are obviously exactly the same.

careful with the conservation of energy when the entire Universe is concerned. According to the Noether theorem, whenever we have a continuous symmetry of the Lagrangian describing the system, there is an associated conservation law. Energy conservation is connected to time shift invariance of physical laws.[5] Einstein pointed out that his General Theory of Relativity did not imply a time shift invariance and therefore there is no law of conservation of energy on a large scale in the Universe. Energy is conserved only *locally.*

1.1.2.2 The second law of thermodynamics

The first law imposes a restriction on all thermodynamic transformations by requiring energy conservation. However, there are many phenomena in nature that do not occur even if energy is conserved. The most important phenomenon in thermodynamics that never occurs is the spontaneous transfer of heat from a cold body to a hot body. Although the first law in principle would allow such transfer, in nature we never observe it. The second law of thermodynamics provides the basis to forbid such transformations. We report here the two most clear statements of the second law:

Clausius: "*There exists no thermodynamic transformation whose sole effect is to transfer heat from a colder reservoir to a warmer reservoir.*"

Kelvin: "*There exists no thermodynamic transformation whose sole effect is to extract heat from a reservoir and to convert that heat entirely into work.*"

The macroscopic definition of entropy is:

$$\Delta S = \int \frac{\delta Q_{rev}}{T} \tag{1.4}$$

where δQ_{rev} is a small *reversible* transfer of heat at a temperature T. Although Q is not a function of state, the entropy S is. The second law states that over time $\Delta S \geq 0$ for an isolated system. The second law of thermodynamics thus also defines the direction of heat transfer from hot bodies to cold bodies and determines the maximum efficiency of heat engines through the Carnot theorem.

1.1.2.3 The third law of thermodynamics

The third law of thermodynamics, also referred to as Nernst-Simon theorem, states that: "*The change in entropy that results from any isothermal reversible transformation of a condensed system approaches zero as the temperature approaches zero,*" or:

[5]Given a Lagrangian L=L(q, $\dot{q}$) a time translation is a coordinate transformation such that $q(t) \rightarrow q(t), \dot{q}(t) \rightarrow \dot{q}(t), t \rightarrow t + \epsilon$. This transformation is a symmetry if and only if L does not depend on t. It can be shown that this symmetry of the Lagrangian implies that $\frac{dH}{dt} = -\frac{\partial L}{\partial t}$, where H is the Hamiltonian of the system which can be identified as the total energy of the system. If L does not depend on t, then H=constant.

$$\lim_{T \to 0} \Delta S = 0 \tag{1.5}$$

One consequence of the Nernst-Simon statement is that the entropy of systems that are in thermal equilibrium must tend to a constant at absolute zero because isothermal transformations do not to change entropy at absolute zero. Planck modified Nernst-Simon statement into: "*As* $T \to 0$ *, the entropy of any system in equilibrium approaches a constant that is independent of all other thermodynamic variables.*" The normal assumption for the constant is to be equal to zero although certain systems might still have a *residual entropy* at absolute zero. A particularly somehow frustrating formulation of the third law states that it is impossible to reach absolute zero in a finite number of transformations.

1.1.2.4 The zero-th law of thermodynamics

The zero-th law of thermodynamics states that systems in thermal equilibrium with each other obey the *transitive* relation. In other terms, if system A is in thermal equilibrium with system B, and system B is in thermal equilibrium with system C, then system A is in thermal equilibrium with system C. Maxwell expressed the zero-th law very efficiently by stating that "all heat is of the same kind." The zero-th law provides the basis for the concept of temperature.

1.1.3 The fundamental thermodynamic relation

We have seen that the first law of thermodynamics 1.3 is a relationship between the internal energy U, the heat Q and the work W. While the differential dU is an exact differential, the two differentials δQ and δW are not exact meaning that their values depend on the path of the transformation. We would like now to express the first law in terms of all exact differentials, i.e., in terms of state functions. Let's start with the first law in differential form:

$$dU = \delta Q + \delta W \tag{1.6}$$

for a reversible process we have that:

$$dS = \frac{\delta Q}{T} \tag{1.7}$$

if we now consider pressure-volume work, we have:

$$\delta W = -PdV \tag{1.8}$$

putting together Eqs. 1.6, 1.7 and 1.8, we have:

$$dU = TdS - PdV \tag{1.9}$$

although Eq. 1.9 has been obtained by considering reversible transformations, it is also valid for irreversible transformations because all the variables involved are state variables (assuming T and P constant).

1.1.4 TdS equations

We can re-arrange Eq. 1.9 as:

$$TdS = dU + PdV \tag{1.10}$$

which is the first "TdS" equation and is valid for isolated systems with reversible exchange of heat and PdV work. Using the definition of enthalpy $H = U + PV$ and its differential $dH = dU + PdV + VdP$, Eq. 1.10 becomes:

$$TdS = dH - VdP \tag{1.11}$$

which is the second "TdS" equation. Let's study an isothermal compression of an ideal gas for which $PV = nRT$. In an isothermal compression of an ideal gas, $dH = 0$ because the internal energy is a function of only temperature T ($U = U(T)$) and the product PV is also only a function of temperature T ($PV = nRT$). It follows that, for an isothermal compression, $TdS = -VdP$. We have:

$$TdS = -VdP = -RT\frac{dP}{P} \tag{1.12}$$

after a simple integration we have:

$$T\Delta S = -RT\ln\frac{P_f}{P_i} \tag{1.13}$$

if the final pressure P_f is bigger than the initial pressure P_i then $\Delta Q = T\Delta S$ is a negative quantity, i.e., heat is extracted from the system. In other words, the increase in pressure obtained by an isothermal compression "squeezes" entropy out of a system. If isothermal compression is followed by isentropic expansion, then cooling is achieved as will be discussed in Section 4.1.

1.1.5 Cooling

A relatively simple technique to achieve cooling is through evaporative cooling. We all have a direct experience of evaporative cooling when, during the heat of the summer, our body produces sweat. The evaporation of water on the skin happens at the cost of heat being removed from the skin and used to give enough kinetic energy to the molecules passing from the liquid phase to the vapour phase. The heat used to evaporate water is called "latent heat of vapourisation." At the microscopic level, at thermal equilibrium between a liquid and its vapour, there is a lot of activity (see Fig. 1.1). The molecules

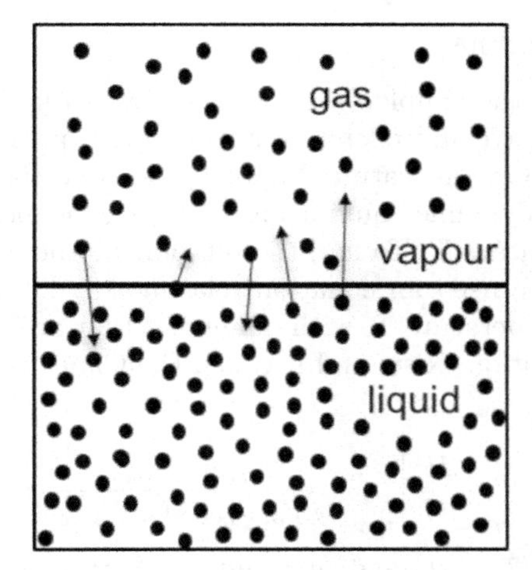

FIGURE 1.1: Liquid in equilibrium with its vapour in a closed system

bound in the liquid and in the vapour (gas) both have a Maxwell distribution of velocities. Those molecules in the liquid, at the high speed tail of the velocity distribution, have enough energy to leave the liquid and pass from liquid into vapour phase. For each molecule evaporated, a small amount of heat is used by the molecule leaving the liquid. At the same time, those molecules in the vapour above the liquid, at the low speed tail of the Maxwell distribution of velocity in the vapour, will be captured by the liquid and pass from vapour into liquid phase. Those molecules captured by the liquid will deposit a small amount of energy. By definition, at the thermodynamic equilibrium of a liquid with its vapour, the energy absorbed by the molecules leaving the liquid to pass into the vapour phase is exactly balanced by the energy deposited by the molecules passing from the vapour into liquid phase.

It is now easy to understand that, if we somehow "unbalance" this equilibrium towards having more molecules evaporating than re-condensing, the energy balance would be in favour of having more energy (heat) removed from the liquid resulting in cooling. A practical way to achieve this unbalance is by pumping over the liquid. Or, as all children know, blow on the soup if you want to cool it. Obviously a much more efficient way to cool the soup would be to put it into a vacuum chamber and pump over it. Such process, although not practical, would certainly amuse the children and cool the soup very quickly.

The following sections deal more rigorously with the processes described above. No attempt is made to provide a comprehensive review of thermodynamics; where this is required, the reader is referred to Refs. [27] and [54].

1.1.6 Vapour pressure

As we have seen in the simple description above, to understand evaporative cooling we need to have an understanding of vapour pressure, latent heat and their dependencies on temperature. Vapour pressure is the pressure exerted by a vapour in thermodynamic equilibrium with its condensed phase in a closed system at a temperature T. To find the relationship between vapour pressure and temperature, we need an equation relating dP/dT, i.e., the variation of pressure P with temperature T, with quantities that are characteristics of the liquid. Such an equation exists and is called the Clausius-Clapeyron equation [66]:

$$\left[\frac{dP}{dT}\right]_{vap} = \frac{S_{gas} - S_{liquid}}{V_{m,gas} - V_{m,liquid}} \tag{1.14}$$

where $S = \Delta Q/T$ is the entropy and V_m is the molar volume. We now identify the quantity $(\Delta Q_{gas} - \Delta Q_{liquid})$ as the latent heat L, and, in addition, we assume that the molar volume of the liquid phase is negligible with the molar volume of the gas phase.

We obtained that the difference in entropies of the liquid and gaseous phases is L/T. Assuming the ideal gas equation, $V_{m,gas} \cong RT/P$, it may be seen that integrating Eq. 1.14 gives

$$P_{vap} \propto e^{-L/RT}. \tag{1.15}$$

Given the logarithmic temperature dependence of Eq. 1.15, it may be considered that, for a closed system in equilibrium, reducing the vapour pressure above the liquid phase (for example, by pumping away some of the vapour) will lead to a spontaneous evaporation, and hence cooling of the liquid.

Notice that the reverse is also the case, in that liquid will be condensed if the pressure is increased and heat is deposited into the liquid.

In the rest of the book, when we talk about *saturated vapour pressure* we mean the pressure of the vapour in equilibrium with its liquid phase, i.e., when there is an equilibrium between the molecules passing from liquid to vapour phase with the molecules passing from the vapour to the liquid phase. The *vapour pressure* is the pressure just above the liquid phase.

In Fig. 1.2, the vapour pressures at low temperatures $< 4.2K$ of the ^{4}He and ^{3}He isotopes are reported. A numeric table is reported in the appendix.

1.1.7 Latent heat

The latent heat of vapourisation (also, the enthalpy of vapourisation) of a substance is the quantity of heat required to convert a given amount of that substance from a liquid to vapour phase *at constant temperature.*

It is important to note that latent heat is also a function of temperature. For a system comprising a liquid with a constant pressure above, the latent heat is equivalent to the amount of energy that is absorbed by the system per

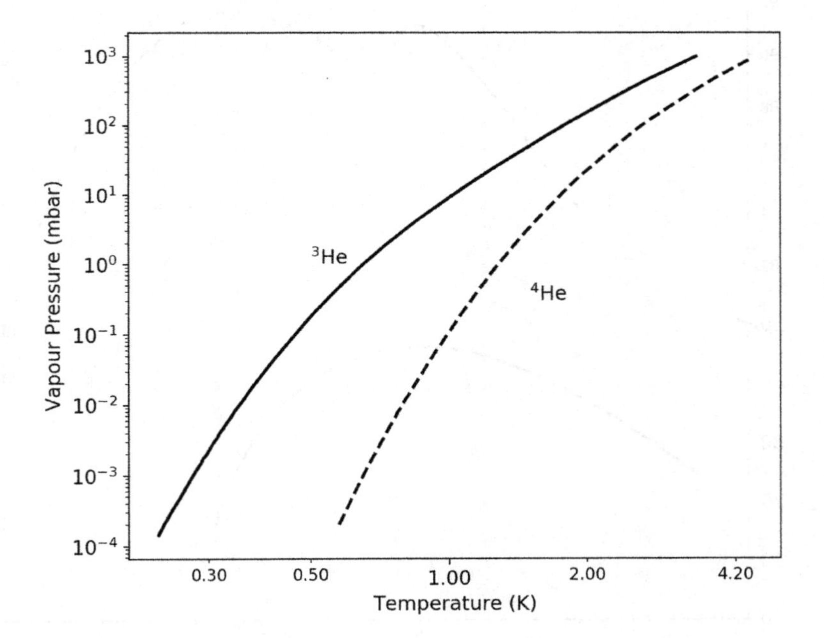

FIGURE 1.2: Vapour pressures of ^{3}He and ^{4}He

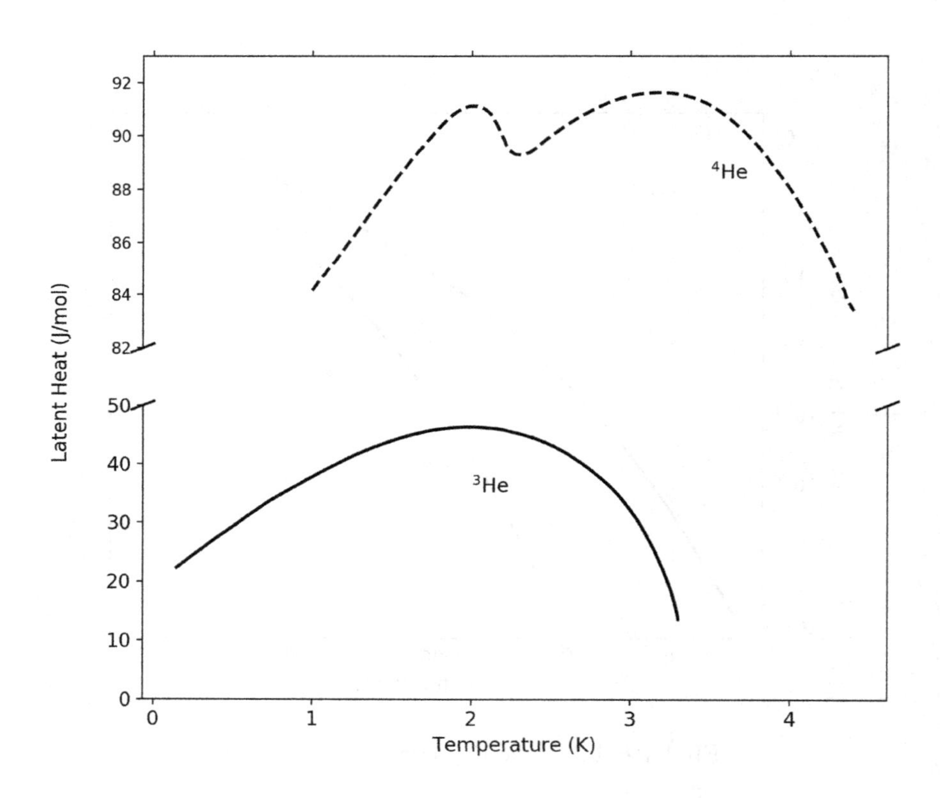

FIGURE 1.3: Latent heats of ^{3}He and ^{4}He

amount of liquid boiled off. Since energy is used to evaporate the liquid, in this case it is negative which means that heat is subtracted from the liquid. In Fig. 1.3, the latent heat of ^{4}He and ^{3}He is shown.

1.1.8 Enthalpy and entropy

In the brief description above, we have used the thermodynamic quantity L, latent heat, and entropy S. We now introduce one more thermodynamic quantity, enthalpy H. In the process of evaporation cooling these entropy, enthalpy and latent heat are all related. We have defined above that the latent heat of vapourisation L is the amount of heat (energy) that must be added to a liquid substance to transform a certain quantity of liquid into vapour. This amount of heat is also termed as *enthalpy of vapourisation*. So formally, we have that $L = \Delta H_{vap}$. The physical units are J/mol.

The enthalpy of vapourisation can be written as:

$$\Delta H_{vap} = \Delta U_{vap} + p\Delta V \tag{1.16}$$

This equation tells us that the change in enthalpy due to evaporation is equal to the change in internal energy when changing from liquid to vapour plus the work done against the ambient pressure. The values of latent heats (or enthalpies) of the various substances therefore tell us something about the energy that keeps the molecules together in the liquid phase. The forces keeping the liquid together are known as Van der Waals forces. These forces are particularly weak for liquid helium (84.5 J/mol) being an inert (noble) gas.

More generally, enthalpy is defined as:

$$H = U + PV \tag{1.17}$$

Eq. 1.17 shows that, for transformation at constant pressure P, the variation of enthalpy is the amount of heat exchanged. The differential of the enthalpy function is:

$$dH = dU + VdP + PdV \tag{1.18}$$

or, using Eq. 1.9:

$$dH = TdS + VdP \tag{1.19}$$

Since Eq. 1.9 involves exclusively state functions, it is valid for all transformations, not just the reversible ones. Eq. 1.9 suggests that the energy in a system is a function of S and V or, in other terms, the energy of a system will not change for those processes at constant volume and entropy. While constant volume is easy to attain experimentally, constant entropy is not easy to achieve. It is therefore useful to introduce other combinations of state functions that can be easier to attain under experimental conditions. For example, another way to describe the evaporation is by identifying the enthalpy of vapourisation as the heat that is absorbed by the surrounding to account for the increase in entropy when a liquid evaporates. To better understand this last point, it is useful to introduce another useful thermodynamic potential: the Gibbs free energy.

1.1.9 Gibbs free energy

We now look for a thermodynamic potential that does not change at constant temperature and pressure, such as for example phase transitions. We define Gibbs free energy with the following combination of state functions:

$$G = H - TS = U + PV - TS \tag{1.20}$$

with the differential given by:

$$dG = -SdT + VdP \tag{1.21}$$

Eq. 1.21 tells us that $dG = 0$ for constant temperature and pressure transformations. In summary, for a vapourisation process, i.e., in a phase transition, we have the following relationships:

$$\Delta H_{vap} = \Delta U_{vap} + p\Delta V \tag{1.22}$$

$$\Delta S_{vap} = S_{vap} - S_{liq} = \Delta H_{vap}/T_b \tag{1.23}$$

$$\Delta G_{vap} = \Delta H_{vap} - T\Delta S_{vap} = 0 \tag{1.24}$$

where T_b is the boiling temperature of the liquid undergoing phase transition. As an example of application of the Gibbs function, let's derive the Clausius-Clapeyron Eq. 1.14. Suppose we have determined the coexistence line on a T, P diagram of two phases that we call phase 1 and phase 2. At any point on the liquid/vapour coexistence curve T, P the specific Gibbs energies of the two phases are unchanged.

We neglect, for now, interface effects which will be treated later. Let's consider a reversible process (T, P constant) in which a little mass dm_1, for example in liquid phase, is transformed into dm_2 (see Fig. 1.4). To calculate the change in G, with $dG = 0$, we write the total Gibbs function as:

$$G = G_1 + G_2 \tag{1.25}$$

or, using the specific Gibbs functions $g_{1,2}$

$$G = m_1 g_1 + m_2 g_2 \tag{1.26}$$

If we take now the differential, for a phase transition we have:

$$dG = g_1 dm_1 + g_2 dm_2 = 0 \tag{1.27}$$

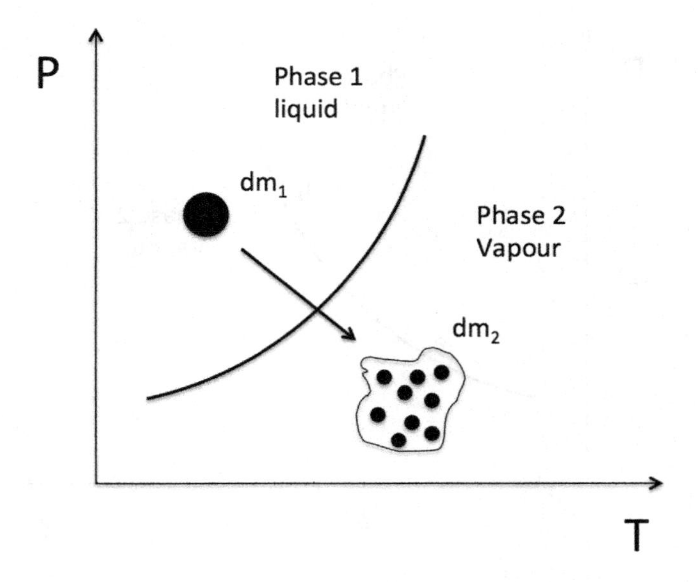

FIGURE 1.4: Two phases T,P diagram

We know that masses are conserved,

$$dm_1 = dm_2 \tag{1.28}$$

or, equivalently

$$g_1 = g_2 \tag{1.29}$$

The Clausius-Clapeyron equation is relating dP/dT to some measurable quantities. With reference to Fig. 1.5, let's consider two points a and b on the coexistence line in the T, P diagram. For these two points, the specific Gibbs functions must satisfy:

$$g_1(a) = g_2(a) \tag{1.30}$$

and

$$g_1(b) = g_2(b) \tag{1.31}$$

in terms of differentials, we have

$$dg_1 = dg_2 \tag{1.32}$$

We already know that the Gibbs function is a function of T, P,

$$g = g(T, P) \tag{1.33}$$

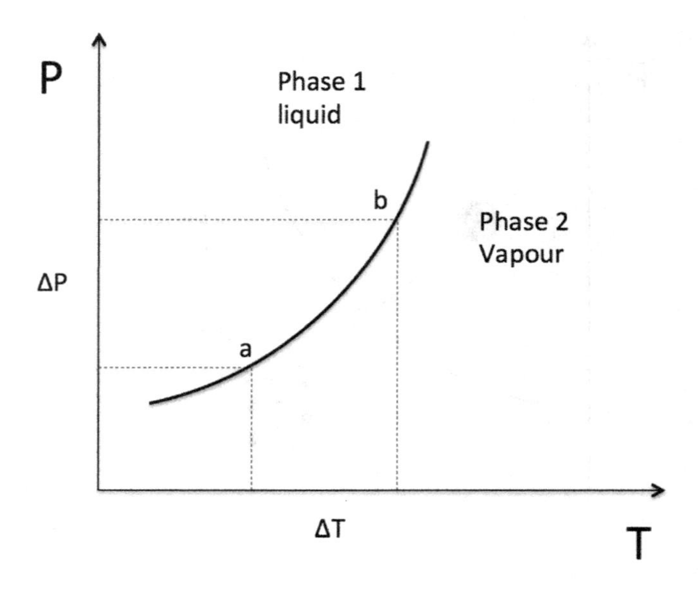

FIGURE 1.5: Two phases ΔT, ΔP relation

taking the differential, we have

$$dg_1 = \left.\frac{\partial g_1}{\partial T}\right|_P dT + \left.\frac{\partial g_1}{\partial P}\right|_T dP = \left.\frac{\partial g_2}{\partial T}\right|_P dT + \left.\frac{\partial g_2}{\partial P}\right|_T dP = dg_2 \tag{1.34}$$

using,

$$dg = vdP - sdT \tag{1.35}$$

by comparison, we see that

$$\left.\frac{\partial g_1}{\partial T}\right|_P = -s \tag{1.36}$$

$$\left.\frac{\partial g_1}{\partial P}\right|_T = v \tag{1.37}$$

using the invariance of the Gibbs function

$$-s_1 dT + v_1 dP = -s_2 dT + v_2 dP \tag{1.38}$$

or,

$$(s_2 - s_1)dT = (v_2 - v_1)dP \tag{1.39}$$

which finally gives the Clausius-Clapeyron equation.

1.1.10 Chemical potential

So far we have considered systems that are allowed to exchange heat and work with the external world. If the systems are also allowed to exchange particles, then we have to incorporate the thermodynamics of this exchange. If we have two systems that can exchange particles, then an equilibrium condition will be reached by properly adjusting the relative concentrations. For example, if one system has more particles than the other, presumably there will be a net flux of particles from the higher to the lower concentrations until equilibrium is reached. If we reason by analogy, when we consider two systems reaching thermal equilibrium, the observable variable of interest is the temperature T. If instead we consider mechanical equilibrium the observable variable is the pressure P. In the case of particle exchange, the associated observable variable is the *chemical potential* μ.

The equilibrium between two systems can be studied by using the entropy variations. In particular, we know that at equilibrium entropy reaches a maximum. If we call N_A and N_B the number of particles, respectively, in the systems A and B that are put in contact and that can exchange particles, we must have:

$$\frac{\partial S_A}{\partial N_A} = \frac{\partial S_B}{\partial N_B} \tag{1.40}$$

It is therefore natural to define the chemical potential as:

$$\mu = -T\frac{\partial S}{\partial N} \tag{1.41}$$

Definition 1.41 tells us that μ has the dimension of energy.[6]

The derivative in Eq. 1.41 is taken keeping all other variables constant. At equilibrium we must have $\mu_A = \mu_B$.

We can re-write the fundamental thermodynamic relation to include the energy associated with the chemical potential. We have:

$$dU = TdS - PdV + \mu dN \tag{1.42}$$

from which we see that μ can be written also as $\mu = \left(\frac{\partial U}{\partial N}\right)_{S,V}$ which gives us an intuitive definition of the chemical potential as the amount of energy added to the system when adding one particle while keeping entropy and volume constant.

In general, when more particle species are present, we have:

$$dU = TdS - PdV + \sum_i \mu_i dN_i \tag{1.43}$$

It is not always possible to keep entropy and volume constants when adding

[6]Note that similar definitions for temperature and pressure are derived in an analog way. In fact, we have $\frac{1}{T} = \left(\frac{\partial S}{\partial U}\right)_V$ and $P = \left(T\frac{\partial S}{\partial V}\right)_U$.

particles. To have a more practical expression, let's write the differential of the Gibbs function $G = U + PV - TS$:

$$dG = dU + VdP + PdV - TdS - SdT \tag{1.44}$$

using 1.43:

$$dG = VdP - SdT + \sum_i \mu_i dN_i \tag{1.45}$$

we therefore see that another useful expression for the chemical potential is:

$$\mu_i = \left(\frac{\partial G}{\partial N_i}\right)_{V,P,N_{j\neq i}} \tag{1.46}$$

which tells us that the chemical potential is the change in Gibbs energy when adding a particle i while keeping constant the volume V, the pressure P and all other particle species $j \neq i$.

1.1.11 Liquid helium

Liquid helium is one of the most important liquids in the field of cryogenics. Entire books have been written about its marvelous properties [95] like, for example, the phenomenon of superfluidity, where liquid helium behaves as a zero viscosity fluid, and the associated fountain effect. A proper understanding of the low-temperature behaviour of liquid helium requires a quantum mechanical treatment. The interested reader can consult various excellent books (see Further Reading at the end of the chapter).

The most common stable helium isotope is ^{4}He, whose nucleus is composed of two neutrons and two protons with antiparallel spins. The total nuclear spin, in its ground state, is therefore equal to zero. The statistical behaviour of large assembly of integer spin systems is described by the Bose-Einstein statistics, and the integer particles are usually referred to as *bosons*.

Helium atoms have two very important characteristics: they are *light* and they are *chemically inert*. As a result, the helium molecules have very little tendency to bound to each other resulting in a very low liquefaction temperature. In addition, being light, the quantum mechanical zero-point energy is non-negligible making helium stay liquid even if we continue reducing the temperature to near zero. We need to apply pressures in excess to 25 bar if we want to solidify ^4He. As we discussed above, boiling point, latent heat of vapourisation and vapour pressure of a liquid are all connected. Liquid helium has a particularly low latent heat mainly because it is the lightest inert gas. The Van der Waals forces between helium atoms are indeed very small due to the absence of static dipole moment. In order to understand the behaviour of helium, we need to quantify the physics of the interaction between the molecules. The force between the molecules depends on the fact that they have a finite size and that they have a mutual interaction. The normal approach is to build an empirical potential describing the so-called *hard core* repulsive interaction and the weaker attractive interaction, also called *London dispersion* mechanism. The hard core repulsive interaction is due to the molecules not being point-like objects, but rather occupying a definite volume. This means that molecules will collide with each other thus explaining the scattering and the mean free path. The way in which the repulsion acts is quantified by the steepness of the potential as the distance between the molecule decreases. This hard core interaction is the dominant term that characterize the deviation of helium from a perfect gas and obviously plays a relevant role at high densities and temperatures. The weaker attraction mechanism can be understood considering that the electrons orbiting around the nucleus of the helium generate an instantaneous electric dipole moment which decreases like $1/r^3$. When two molecules are close, the total potential will be proportional to $1/r^6$. Several potentials have been proposed to describe the helium interactions. The Lennard-Jones potential is one of the most popular and is defined as:

$$V_{LJ}(r) = 4\epsilon_0 \left[\left(\frac{r_0}{r}\right)^{12} - \left(\frac{r_0}{r}\right)^{6} \right] \tag{1.47}$$

where ϵ_0 is related to the minimum depth of the potential and r_0 give the scale of the dimension of the molecule. Good values for helium are $r_0 =$ 2.556 Angstrom and $\epsilon_0 = 10.22\ k_B$ where k_B is the Boltzmann constant. The potential is shown in Fig. 1.6.

The quantum mechanical zero-point energy E_0, being proportional to $1/m$, produces relatively big amplitudes of vibrations thus explaining the low latent heat of vapourisation. The zero point energy can be espressed in terms of the Planck's constant h and the atomic density N/V:

$$E_0 \simeq \frac{h^2}{8m}\left(\frac{N}{V}\right)^{2/3} \tag{1.48}$$

The low heat of vapourisation means that liquid helium has a low cooling power. However, the fact that it stays liquid at very low temperatures with relatively high vapour pressure will allow to produce additional cooling through evaporation cooling. The rarer helium isotope ^{3}He has a nucleus with 2 protons and only one neutron. Having lost a spin-1/2 particle, the ^{3}He atoms obey the Fermi-Dirac statistics or, in other terms, is a *fermion.*

The different statistical properties of the two isotopes give rise to substantially different low-temperature characteristics. First of all, ^{4}He is subject to Bose-Einstein condensation while ^{3}He is subject to Fermi-Dirac condensation. Secondly, having one less neutron, ^{3}He is lighter and therefore its quantum mechanical zero-point energy is even larger than ^{4}He explaining, for example, the lower boiling temperature (3.2 K compared with 4.2 K for ^{4}He), smaller latent heat and larger vapour pressure. The larger vapour pressure, in particular, makes a liquid ^{3}He pumped bath reach much lower temperatures than a

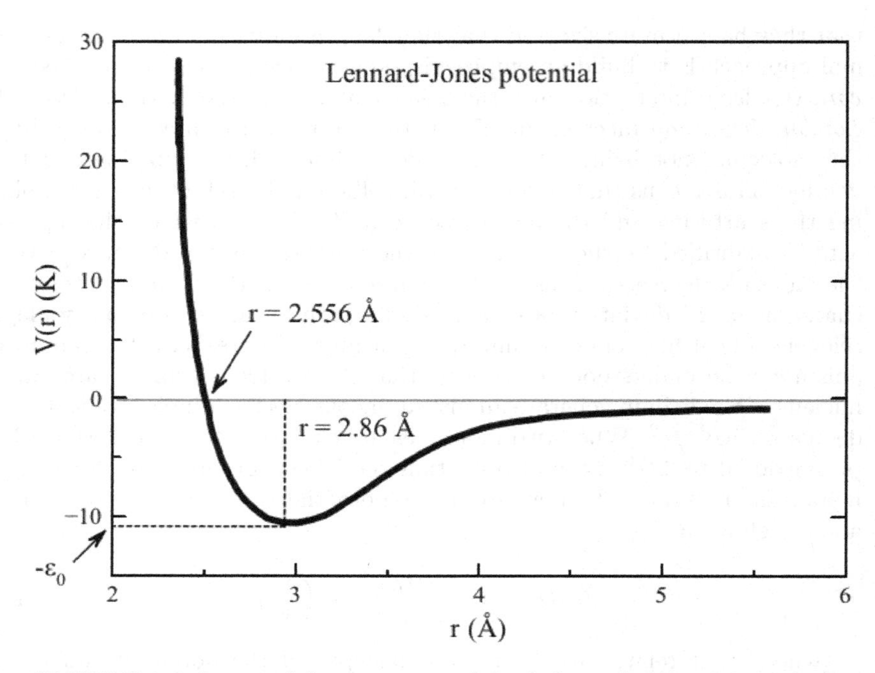

FIGURE 1.6: Lennard-Jones potential for helium (adapted from [28])

pumped ^{4}He bath (0.3 K for ^{3}He compared to 1 K for ^{4}He). The main downside of the usage of ^{3}He is its extremely high cost due to its extremely low natural abundance.

As shown in Table 1.1, ^{4}He has a relatively high superfluid transition. Below this temperature, the ^{4}He has two components, one normal and one superfluid. The last one becomes the dominant component when the temperature continues to decrease. The presence of this superfluid component changes completely the behaviour of the liquid and consequentially also its properties change dramatically. One of the most important changes is the difference in the thermal conductivity of the liquid. In Fig. 1.7, it is possible to notice that for normal ^{4}He the thermal conductivity is of the order of mW/cm/K. Instead in case of superfluid, the thermal conductivity is several orders of magnitude bigger as can be seen by comparing Fig. 1.7 with Fig. 1.8. More properties of both helium isotopes are presented in Appendix A.

1.2 QUANTUM EFFECTS AT LOW TEMPERATURE

The field of low-temperature physics has the "privilege" to observe quantum phenomena at macroscopic scale. In particular, we have *superconductivity* and *superfluidity* that are both described by quantum mechanics. A complete thorough treatment of superfluids and superconductors is beyond the scope of this

	^3He	^4He
Boiling Point T_b (K)	3.19	4.21
Critical Temperature T_c (K)	3.32	5.20
Maximum Superfluid Transition Temperature T_c (K)	0.0025	2.1768
Density[a] ρ (g cm^{-3})	0.082	0.1451
Classical Molar Volume[a] V_m (cm^3 mol^{-1})	12	12
Actual Molar Volume[a] V_m (cm^3 mol^{-1})	36.84	27.58
Melting Pressure[b] P_m (bar)	34.29	25.36
Minimum Melting Pressure P_m (bar)	29.31	25.33
Gas-to-Liquid Volume Ratio[c]	662	866

From [66].
[a] At saturated vapor pressure and $T = 0$ K.
[b] At $T = 0$ K.
[c] Liquid at 1 K and NTP gas, $T = 300$ K and $P = 1$ bar.

TABLE 1.1: Properties of liquid helium isotopes

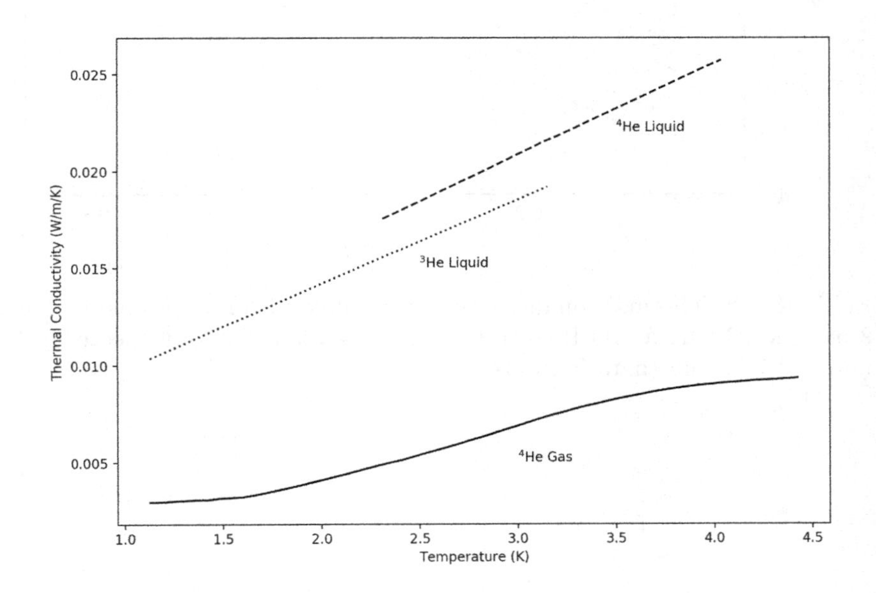

FIGURE 1.7: Thermal conductivity of the two helium isotopes below 4 K (data from [49])

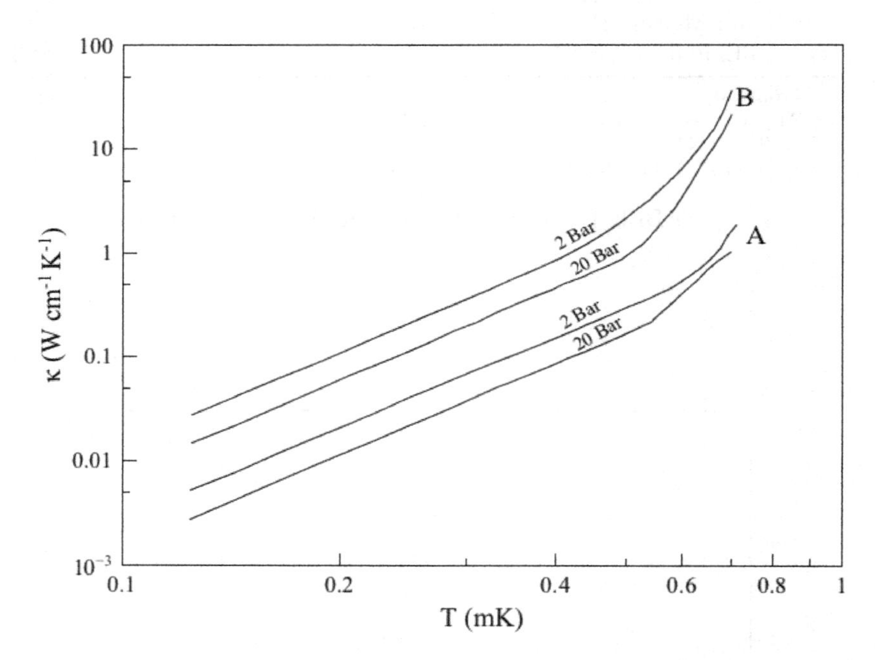

FIGURE 1.8: Thermal conductivity of superfluid ^{4}He isotopes below 1 K at 2 bar and 20 bar; A and B correspond to tube diameters of respectively 1.38 mm and 7.97 mm (data from [32])

book and we suggest the interested reader to consult various excellent books (see Further Reading at the end of the chapter).

1.2.1 Superconductivity

Superconductivity was first observed by Heike Kamerlingh Onnes in 1911 and consists of exactly zero electrical resistance in a material when cooled below a critical temperature T_c. A theory of superconductivity was first proposed by Bardeen, Cooper and Schrieffer in 1957 [6] and is commonly referred to as BCS theory. For $T < T_c$, electrons do not act as a single particles, but they gather together in pairs called *Cooper pairs*. A pair is not strongly bound and there is a probability to break it that is proportional to $exp(-E_{couple}/kT)$ according to Boltzmann theory. As we know, the electrons are half-integer spin particles (*fermions*) so they follow the Fermi-Dirac statistics. When a Cooper pair is formed it acts like a boson: when the temperature is reduced, the pairs are not limited by the Pauli exclusion principle and can reach, and share, the lowest energy level. They form a so-called Bose condensate. Another remarkable behaviour consists of the complete expulsion of the magnetic flux fields from within the superconductor (*Meissner effect*).

The behaviour of the magnetic field inside a superconductor is governed by the (second) London equation:

$$\nabla^2 \mathbf{B} = \frac{1}{\lambda^2}\mathbf{B} \tag{1.49}$$

where $\mathbf{B}$ is the magnetic field and λ is called the *London penetration depth* and is defined as:

$$\lambda^2 = \frac{m}{\mu_0 n e^2} \tag{1.50}$$

where m, μ_0, n and e are, respectively, the mass of the superconducting charge carriers, the vacuum permeability, the charge carrier density and the charge of the carriers. In a simple representative case of a superconductor immersed in a constant magnetic field along the z axis, for a direction x perpendicular to z, the solution to the London equation is of the form:

$$B(x) = B_0 e^{-x/\lambda} \tag{1.51}$$

and predicts that the field inside a superconductor decays exponentially. Measurements on superconducting metals have shown that the heat capacity has a discontinuity at the transition temperature. At low temperatures, the heat capacity of metals is given by:

$$C = \gamma T + AT^3 \tag{1.52}$$

where the linear term proportional to T is due to the conduction electrons while the term proportional to T^3 is due to the phonons and is related to the lattice heat capacity. It is an experimental fact that the phonon term is Eq. 1.52 does not contribute to the heat capacity discontinuity. It follows that the discontinuity must be due to the ability of the conduction electrons to absorb heat. It is therefore natural to think of the superconductor as containing two *fluids*: (1) normal electrons and (2) superconducting electrons (Cooper pairs). The thermal conductivity in a normal metal can be described by the following equation:

$$\sigma_n = \sigma_{el,n} + \sigma_{ph,n} \tag{1.53}$$

In the superconducting state, the thermal conductivity can be written as:

$$\sigma_s = \sigma_{el,s} + \sigma_{ph,s} \tag{1.54}$$

where the subscripts n and s indicate, respectively, the normal and superconductive state, *el* the electronic contribution to the thermal conductivity and *ph* the phonon contribution. The Cooper pairs, carrying the supercurrent, are not expected to contribute to the thermal current. It follows that the electron term $\sigma_{el,s}$ must be less than the term $\sigma_{el,n}$. Therefore, since the heat is transported only by the fraction of normal electrons, we expect that the thermal conductivity of a metal, below the transition temperature, is less than the thermal conductivity in the normal state.

Above T_c the electrons are the only conductor of electricity and heat; below T_c normal electrons and Cooper pairs coexist. With the decrease in temperature below T_c, more and more electrons pair into Cooper pairs, becoming bosons and condense into the superconducting state. At $T = 0$ all the electrons are coupled into Cooper pairs and no normal electrons are found. It follows that we always have that:

$$\sigma_n > \sigma_s \tag{1.55}$$

According to this simple model, when heat is applied to a superconductor, part of the heat goes to the lattice and to the normal electrons, while part of the heat goes to break some Cooper pairs since the superconducting state is a lower energy state. The discontinuity is therefore due to the normal/superconducting electron spectrum. Once the lattice contribution is subtracted, the residual electron portion of the heat capacity can be modeled by the function:

$$C_{el} = Ae^{-D/T} \tag{1.56}$$

which indicate the existence of an energy gap. Cooper pairs lie in the lower energy state, and to be excited to normal electron they need to overcome the superconducting gap.

In conclusion, the thermal conductivity of metals capable of becoming superconductors depends on the temperature. We have seen that Cooper pairs neither accept heat (less than the energy gap) nor scatter phonons. For those metals in which the thermal conduction is due mostly to the electrons, the thermal conductivity below T_c decreases with temperature as normal electrons condense into Cooper pairs.

It is important to consider that for those metals in highly disordered states or with impurities, the thermal conduction is dominated by phonons and thermal conductivity can increase with temperature below T_c. In this case, there are less normal electrons around to scatter phonons.

1.2.2 Superfluid helium

Another remarkable quantum mechanical phenomenon happening at low temperatures is the complete disappearance of any viscosity in liquid ^{4}He when cooled below a transition temperature of $T_\lambda = 2.1768$ K. In Fig. 1.9a, the phase diagram of ^{4}He is shown. Two striking anomalies are immediately visible: the liquid phase extends all the way down to $T = 0$ K and only by applying a pressure in excess of 25 bar will make solid helium; there are two distinct liquid phases: namely He-I and He-II. He-I is normal fluid while He-II is superfluid, i.e., frictionless flow. The existence of liquid with approaching zero temperature is a quantum effect due to the inert nature of helium and its low mass. Because of its low mass, zero-point fluctuations are large enough to prevent the molecules to stick and make a solid (unless $P > 25$ bar). As in the case of superconductivity, also in the case of superfluid helium, the study of its heat capacity will shed some light on the anomalous phenomena. In Fig. 1.3, the latent heat of liquid ^{4}He is shown plotted versus temperature. The kind of anomalous behaviour shown around 2.2 K in Fig. 1.3 is usually associated with order-disorder transitions. The liquid must be in some completely ordered state at $T = 0$ of quantum mechanical origin (the entropy should vanish at $T = 0$). It is supposed that the superfluidity is due to this ordering among quantum mechanical energy levels instead of position.

In analogy with the case of superconductivity discussed above, a two-fluids model can be attempted to explain the peculiar behaviour of liquid ^{4}He. According to this model, the superfluid phase is described by a mixture of two components: a normal component characterized by its density ρ_n, its velocity v_n and its viscosity η_n, and a superfluid component with associated density ρ_s and velocity v_s. Above T_λ the superfluid component is absent while below about $T = 1$ K the normal component is practically negligible. In the temperature interval 1 K $< T <$ 2.17 K the two components coexist. Since an increase in temperature increases ρ_n and decreases ρ_s a temperature gradient will drive the superfluid components towards higher temperatures. At the same time, the normal fluid component will be driven in the opposite direction.

A pressure gradient will tend to drive both fluids in the same direction. An increase in temperature increases ρ_n but decreases ρ_s, so a temperature gradient tends to drive the superfluid component in one direction (towards higher temperature) and the normal fluid in the opposite direction. Therefore, in superfluid helium, heat transport takes place by counterflow of the two

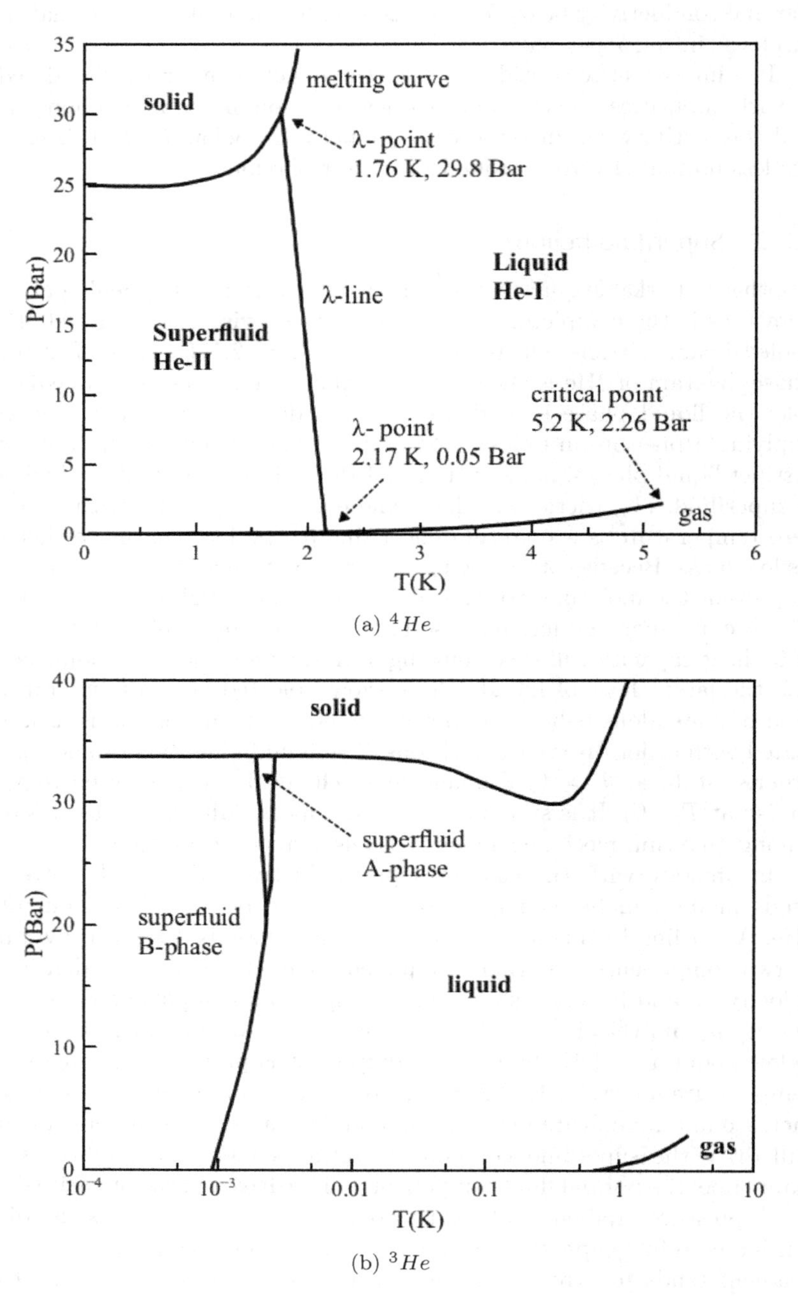

(a) ^4He

(b) ^3He

FIGURE 1.9: Phase diagram of the two stable helium isotopes (data from [91])

components. However, *only the normal component carries heat* at a rate per unit area given by:

$$Q = \rho_n S T v_n \tag{1.57}$$

where S is the entropy per unit mass of helium. This mechanism provides a very effective heat conduction mechanism.

The equation describing the acceleration of the superfluid component [95] is

$$\frac{\partial v_s}{\partial t} = -\frac{1}{\rho}\nabla P + s\nabla T. \tag{1.58}$$

and it may hence be seen that (a) the superfluid flows under the influence of pressure gradients and (b) that it also responds to gradients in temperature. From the so-called "fountain" term ($s\nabla T$) we see that the superfluid will accelerate towards warmer regions giving rise to the thermomechanical or "fountain pump" effect: if an empty beaker is lowered into a bath of Helium-II, then a liquid film develops over the surface of the container until the levels are equalized [95]. Similarly, a pot of Helium-II will see a superfluid film develop up the walls to coat the inside of the pot.

It may be seen therefore how for ^{4}He coolers, superfluidity introduces additional complications. As a film will tend to climb the inside of the pot, liquid may be drawn out of the pot and into the pumping tube where (a) it evaporates without producing useful cooling and (b) the presence of the film increases the thermal load on the evaporator.

1.3 VACUUM AND GAS

Cryogenic systems are always associated with vacuum systems, unless the system is in space where high vacuum is a natural occurrence (intergalactic space can have a vacuum as low as $10^{-18} hPa$). In a normal laboratory environment, we have air at approximately 1 bar at a temperature of 300 K. It is a common experience that if we expose a system to cold air, several atmospheric gas and vapour constituents will condense on the system. The reason is that heat tends to flow from high temperature to low temperature through a thermal conductive medium — in this case the air molecules. The net effect will be that your cold system will tend to heat up at the expenses of a cooling of the surrounding air. If the system is supposed to be operated at cryogenic temperatures, it is absolutely essential that we shield it from direct contact with air thus the usage of so-called *cryostats*. Cryostats are usually metal vessels where the cryogenic system is placed and shielded from thermal inputs coming from radiation, conduction and convection of the 300 K ambient temperature. It is useful to introduce a nomenclature of the various kind of vacuum (see Table 1.2).

Considering that 1 hPa = 1 mbar, we see from Table 1.2 that common

Pressure	hPa	Mean Free Path (m)
Atmosphere	$1,013.25$	$6.8 \cdot 10^{-8}$
Low Vacuum (LV)	$300 \ldots 1$	$10^{-8} \ldots 10^{-4}$
Medium Vacuum (MV)	$1 \ldots 10^{-3}$	$10^{-4} \ldots 10^{-1}$
High Vacuum (HV)	$10^{-3} \ldots 10^{-7}$	$10^{-1} \ldots 10^{3}$
Ultra-High Vacuum (UHV)	$10^{-7} \ldots 10^{-12}$	$10^{3} \ldots 10^{8}$

TABLE 1.2: Vacuum pressure regimes

mechanical roughing pumps operate in the Medium Vacuum range, while diffusion and turbo pumps operate in the High Vacuum range.

1.3.1 Gas equations

In vacuum technology we will deal mostly with gas. It is therefore useful to briefly discuss the gas equations that are relevant when encountering vacuum systems. The definition of *normal condition* for a gas is the following: a gas at 101,325 Pa and at a temperature of 273.15 K. Under normal conditions, one mole of gas will occupy a volume of 22.414 litres. Remember that one mole of any material contains exactly the Avogadro number $N_A = 6.022 \cdot 10^{23}$ of particles (atoms or molecules). 1 mole is defined as the amount of substance that contains the same number of particles as the number of atoms of 12 g of ^{12}C.

The simplest equation relating pressure, volume and temperature of an *ideal* gas is

$$PV = nRT = NkT \tag{1.59}$$

where n is the number of moles of gas, P is the pressure (Pa), V is the volume (m^3), T (K) is the temperature, R is the gas constant ($R = 8.3144598(48) J \cdot mol^{-1} \cdot K^{-1}$), N is the number of molecules and k is the Boltzmann constant ($k = 1.38064852 \times 10^{-23} m^2 \cdot kg \cdot s^{-2} \cdot K^{-1}$). This equation has a good range of validity and can be used for gases that are far from critical conditions, like critical pressure or condensation temperature. For Helium, for example, Eq. 1.59 is valid to within 10% for temperatures as low as 8 K. In order to account for non-ideal conditions, terms can be added to Eq. 1.59:

$$\frac{PV}{nRT} = Z = 1 + B(T)\frac{n}{V} + C(T)\frac{n^2}{V^2} + D(T)\frac{n^3}{V^3} + \ldots \tag{1.60}$$

Eq. 1.60 is called *virial expansion* and the coefficients $B(T), C(T), D(T)$ etc. are called, respectively, second coefficient, third coefficient, fourth coefficient, and so on. Z is the *compression factor* and gives an indication of the deviation from ideality of the gas.

The speed of the various molecules in a gas obeys the Maxwell-Boltzmann distribution:

$$f(v) = \sqrt{(\frac{m}{2\pi kT})^3}\, 4\pi v^2\, e^{-\frac{mv^2}{2kT}} \tag{1.61}$$

The maximum of the distribution of Eq. 1.61 is the *most probable* speed v_{mp} among the molecules which is different from the average speed v_{av} or rms speed v_{rms}. We have that:

$$v_{mp} = \sqrt{\frac{2RT}{M}} < v_{av} = \sqrt{\frac{8RT}{\pi M}} < v_{mp} = \sqrt{\frac{3RT}{M}} \tag{1.62}$$

1.3.2 Mean free path

In a gas, the mean free path is the *average* distance that a molecule travels before colliding with another molecule. Like speed in a gas, it is a statistical quantity with an associated distribution function. Assuming the molecules follow the Maxwell-Boltzmann distribution, the mean free path is defined as:

$$\ell = \frac{1}{\sqrt{2} n\sigma} \tag{1.63}$$

where $n = N/V$ is the number of molecules per cubic metre and sigma is the cross-sectional area for collision. In a simple classical case, assuming ideal gas of hard spherical molecules, if d is the diameter of the spherical molecule, Eq. 1.63 becomes:

$$\ell_{cl} = \frac{k_B T}{\sqrt{2}\pi d^2 P} \tag{1.64}$$

in a more realistic approximation, Lennard-Jones potentials need to be used and we have:

$$\ell_{LJ} = \frac{\mu}{P}\sqrt{\frac{\pi k_B T}{2m}} \tag{1.65}$$

where μ is the viscosity and m is the mass of the molecule.

1.3.3 Transport properties: heat transfer of ideal gas

Transport properties of a substance, in our case a gas, refer to the transport of energy, matter and momentum from one point in space to another. In order to have transport, a gradient must be present in such a way to generate a *force.* If one point of the system is hotter than another point, then the motion of the molecules will statistically transport energy from the hot point to the cold point: this is thermal conduction. If a solid is moving at a certain speed within a fluid, then the molecules close to the solid will acquire momentum by collision and this momentum is transferred to more distant molecules.

Transport of momentum is related to viscosity. Finally, if a portion of a fluid has higher concentration than the rest of the fluid, the mass will distribute in such a way to equalize the density by moving mass around. There is a mass movement called diffusion.

Each of these transport processes are ruled by a *transport equation* which relates a gradient of a quantity to a generic flux, i.e., the amount of transported quantity per unit time, unit area. In other terms, to move something we need a force which is the result of a gradient. We report here the simple case of thermal conductivity of an ideal gas to show some remarkable features that are useful when treating cryogenic systems. More detailed accurate calculations for real gases are beyond the scope of this book.

The generic gradient equation proportional to the flux, in the case of heat conduction, is called *Fourier's Law*:

$$\frac{dQ}{dT} = -\kappa A \frac{dT}{dx} \tag{1.66}$$

and relates the heat flux $\frac{dQ}{dT}$ to the temperature gradient $\frac{dT}{dx}$. A is the area and κ is the proportionality factor (units $Wm^{-1}K^{-1}$) expressing how well a substance will transport heat. The minus sign in Eq. 1.66 simply describes the tendency for heat to go from hot to cold bodies.

Using the theory of kinetic energy in an ideal gas, it can be shown that the heat conduction Eq. 1.66 becomes:

$$\frac{dQ}{dT} = -\frac{1}{3} C v_{av} \ell A \frac{dT}{dx} \tag{1.67}$$

where C is the specific heat of ideal gas. If we compare Eq. 1.66 with 1.67 we have an expression of the thermal conductivity of an ideal gas:

$$\kappa = -\frac{1}{3} C v_{av} \ell \tag{1.68}$$

Eq. 1.68 shows that the thermal conductivity *does not* depend on the pressure of the gas. So what is the point in evacuating a cryostat? As we decrease the pressure, the mean free path increases. When the mean free path of the residual gas is of the order of the size of the vacuum vessel, any decrease in the pressure does not modify the mean free path being limited by the size of the vessel. In this regime, the heat capacity is now dominating the thermal conductivity.

1.4 MATERIAL PROPERTIES AT LOW TEMPERATURE

When designing low-temperature refrigerators, in particular closed-cycle sorption coolers, we need to put special care in the selection of materials used. There are two classes of metals that are commonly used: high thermal conductivity metals (usually copper but also aluminium) and low thermal conductivity metals (usually stainless steel but other metals are possible like, for example, titanium). High-conductivity metals are used wherever isothermal conditions are required between the cold liquid and the mass to be cooled. Low-conductivity metals are used wherever different temperature stages inside an apparatus need to be connected structurally while minimizing the thermal conduction.

In order to characterize the materials and especially the solids at low temperature, we need to divide the properties in two different categories: *mechanical properties* and *thermal properties*.

1.4.1 Mechanical properties

This category refers to all the properties that describe the mechanical (static or dynamical) behavior of a material. Mechanical properties also include the *elastic modulus* and *Poisson ratio*. The first quantity describes the resistance of any material when it is deformed non-permanently by an external stress and is defined as

$$E = \frac{\sigma}{\epsilon} \tag{1.69}$$

where σ is the *stress*, defined as F/A, and ϵ is the *strain* defined as $\Delta L/L$. The strain should not be confused with the linear expansion. Indeed, the strain is induced by a force that causes the contraction.

Instead, the second one is given by the negative of the ratio between the transverse and axial strain as

$$\nu = -\frac{d\epsilon_t}{d\epsilon_a} \tag{1.70}$$

where the subscript t and a mean the transverse component and the axial component of the strain.

Both properties are strongly temperature dependent. However, it is important to underline that usually the elastic modulus increases when the temperature decreases (as shown in Fig. 1.10). This is important because it means that usually an object that is designed to be mechanically resistant at room temperature will be resistant also at low temperatures.

These properties are particularly useful in designing a cryostat, since they will constrain the dimensions of a cryostat due to the material properties limit. Indeed, inside the cryostat the pressure is usually six orders of magnitude below the atmospheric pressure, this means that there will be a huge force applied on the external shell of the cryostat. The deflection due to the differential pressure can be important and needs to be studied accurately. In fact, if the cryostat has some lens or windows to separate the internal environment with respect to the external, the deflection can cause a significant movement of the lens or the windows and so causing a new optical design.

Another important application of these quantities in cryogenics is related to the design of sorption coolers. These components can contain up to 90 bar

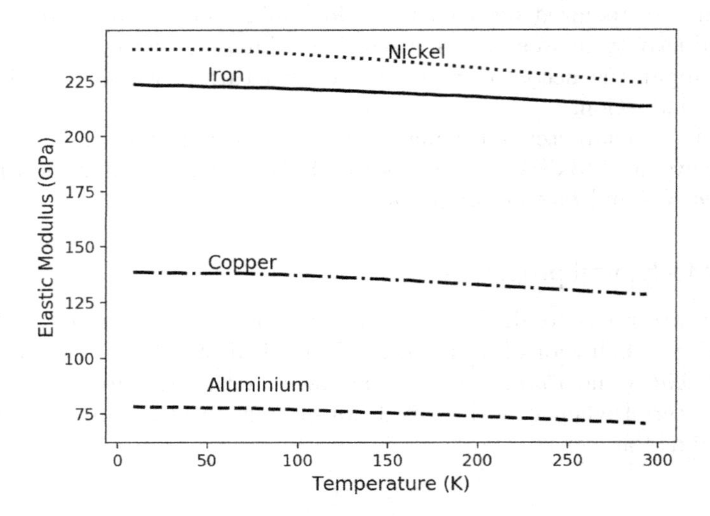

FIGURE 1.10: Elastic modulus of some materials from room temperature to cryogenic (data from [48])

at room temperature of helium, so they need to be designed in order to bear the differential pressure.

1.4.2 Thermal properties

This category refers to all the properties that describe how a material reacts to applied heat. This one can be applied directly or pass through the material.

Despite the importance of the previous category, this one is the most important when we are dealing with sorption cooler. In particular the interesting properties are *thermal conductivity*, *specific heat* and *linear expansion*. Thermal conductivity is the property that describes the ability of a material to transfer heat between two points via conduction. Specific heat capacity is the quantity of energy required to raise the temperature of a material by 1 degree. The coefficient of linear expansion describes the change in dimension of the material due to a change in temperature.

These properties have a more direct applications in designing sorption coolers. As we will describe in detail in Chapter 4, sorption coolers need to be accurately designed to reduce the parasitic heat load due to the tubes that connect the coldest part with the other components. This means that for the tubes a particular material with low thermal conductivity needs to be chosen. However, there is the necessity to consider that the tubes will shrink during the cooldown of the sorption cooler, so the length of the tubes at room temperature needs to consider the effect of the thermal contraction. The relevance of the

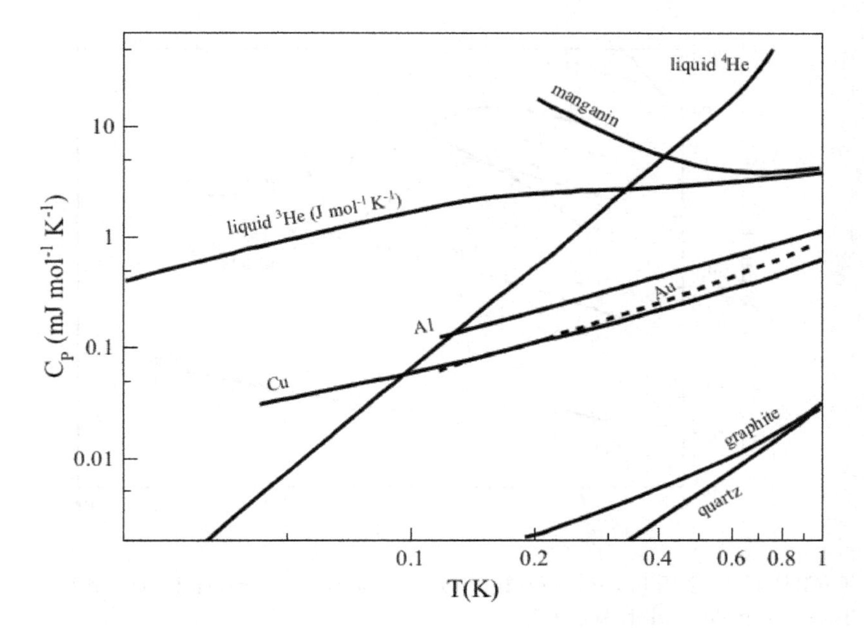

FIGURE 1.11: Specific heat of some materials below 1 K (data from [52])

specific heat in designing a sorption cooler is the consideration of the quantity of heat that must be removed in order to cool cycling components from 50 K to 4 K. As we will see in Chapter 4, this will prove to be significant.

As for mechanical properties, also the thermal ones are strongly temperature dependent, but some common behaviours can be traced. For example for the metals, the thermal conductivity tends to have a peak between 10 K and 50 K and decreases quite quickly below 10 K. Different properties for different materials are shown in Figs. 1.11—1.13.

The most important materials used in cryogenics applications are stainless steel, aluminium and copper. Each of these has several alloys that can have significantly different behaviour at low temperature. However, in general, stainless steel is used due to its low thermal conductivity and its high mechanical resistance. Also, copper maintains high thermal conductivity, but there is the necessity to consider that it has a density which is quite high. Finally, aluminium has a thermal conductivity inferior to copper but it has a density that is inferior too, which means less weight to sustain for the support.

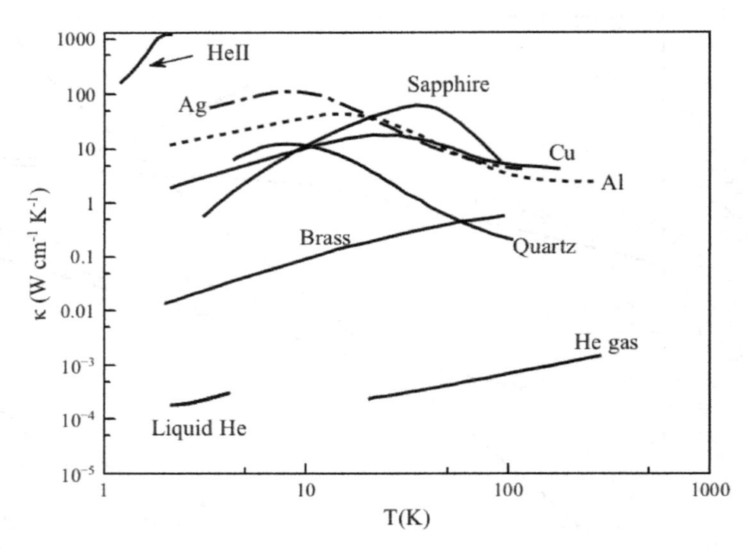

FIGURE 1.12: Thermal conductivity of selected materials above 2 K; for a more complete plot, see [66]

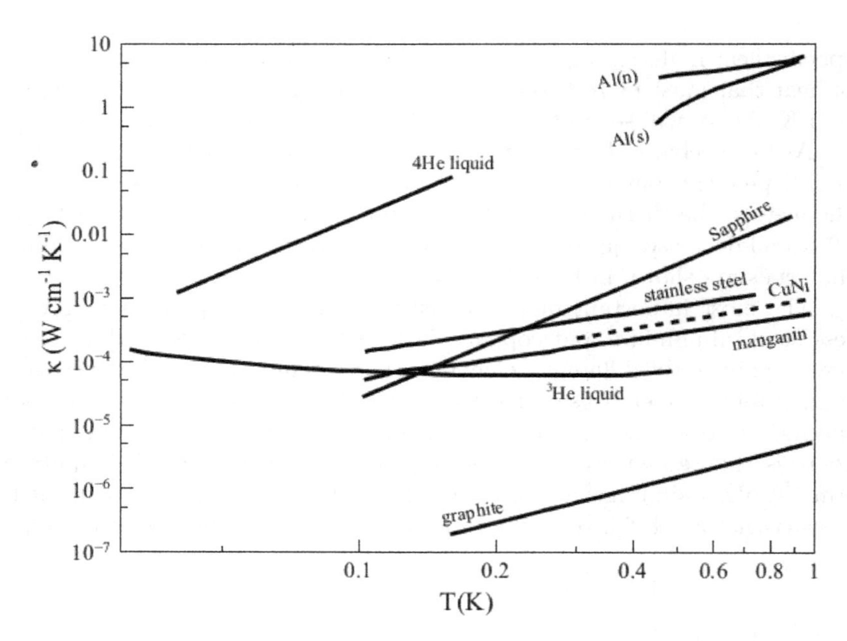

FIGURE 1.13: Thermal conductivity of a selection of different materials below 1 K (data from [52])

FURTHER READING

London, F. (1954). *Superfluids.* Wiley.

Lounasmaa, O.V. (1974). *Experimental Principles and Methods Below 1 K.* Academic Press

Mandl, F. (1988). *Statistical Physics.* Wiley.

Mendelssohn, K. (1966). *The Quest for Absolute Zero: The Meaning of Low Temperature Physics.* McGraw-Hill.

Pobell, F. (2007). *Matter and Methods at Low Temperatures.* Springer.

Tilley, D.R. and Tilley, J. (1990). *Superfluidity and Superconductivity.* CRC Press.

Van Sciver, S.W. (2012). *Helium Cryogenics.* Springer.

Wilks, J. and Betts, D. (1987). *An Introduction to Liquid Helium.* Clarendon Press.

CHAPTER 2

Heat Transfer

HEAT TRANSFER is probably the most important process in thermodynamics. In cryogenics it is also of paramount importance since cooling must happen by removing heat from a fluid or a body. In this chapter we will cover the mechanisms of heat transfer, i.e., conduction, convection and radiation. In practically every cryogenic system, all these mechanisms need to be evaluated either to be minimized, if we want to achieve thermal isolation, or to be maximized, if we want to achieve good thermal contact.

2.1 MECHANISMS OF HEAT TRANSFER

2.1.1 Conduction

conduction describes the heat transfer through a material due to collisions of the particles. For this reason, the heat transferred through this mechanism is exclusively confined inside the body. In Chapter 1 we have already treated the simple case of heat transfer in an ideal gas through the use of the Fourier's equation. We now re-write it in a slightly different notation:

$$\vec{q} = -k\nabla T \tag{2.1}$$

where $\vec{q}$ is the heat flux through a surface in W/m^2, T is the temperature distribution in the material which can be dependent on the position and finally k is the *thermal conductivity*. This term is a property of the material, is dependent on the temperature and, in case the material is not isotropic, is dependent also on the position so $k = k(T, \vec{r})$. However, in many cryogenic applications, the materials used are considered isotropic because there is no evidence of non-isotropic behaviour. In this way, the dependence on the position disappears while the temperature dependence remains.

Usually, there is no need for solving Eq. 2.1 in three dimensions and it is enough to solve it in one dimension. In this case we can re-write Eq. 2.1 as:

$$q = -k\frac{dT}{dx} \tag{2.2}$$

In the simple case of thermal conductivity not depending on temperature, it is possible to write Eq. 2.2 as:

$$qdx = -kdT \tag{2.3}$$

which can be simply integrated. In many applications, the heat flux is not important, instead it is more important that the heat is transferred, so $Q = Aq$ where A is the cross-sectional area and q is the heat transferred per unit area. We can therefore write Eq. 2.3 as:

$$\frac{Q}{A}dx = -kdT \tag{2.4}$$

so the total heat transferred is:

$$Q = \frac{-\int_{hot}^{cold} kdT}{\int_{x2}^{x1} \frac{dx}{A}} \tag{2.5}$$

Using the mean value integral theorem, it is possible to write the numerator as:

$$-\int_{hot}^{cold} kdT = k_m(T_h - T_c) = k_m \Delta T \tag{2.6}$$

Combining Eq. 2.5 and Eq. 2.6, it is possible to write:

$$Q = \frac{\Delta T}{\frac{1}{k_m}\int_{x2}^{x1} \frac{dx}{A}} = \frac{\Delta T}{R_{cond}} \tag{2.7}$$

In the last paragraph, we introduced thermal resistance R_T. This term allows us to find an analogy between Fourier's equation in one dimension and Ohm's law. In particular, it is possible to consider Q as the thermodynamical equivalent of the electric current I, ΔT as the equivalent of the voltage ΔV and R_T as the electrical resistance. With this analogy, it is possible to apply all the rules for the resistor electrical circuit to heat transfer, so every component behaves as a resistance in circuit between two points with different temperatures. This analogy helps especially when it is requested to compute the conduction between a composite material. Indeed, this can be simplified as a series and/or parallels of different thermal resistance R_T, so the equivalent thermal resistance can be computed following the standard rules for electric resistances. Every interface between components or materials needs also to take into consideration that there is a discontinuity between the two different materials and this creates a contact resistance which can be modeled as:

$$R_{Tc} = \frac{1}{h_c A} \tag{2.8}$$

where h_c is the contact heat transfer coefficient. The form of Eq. 2.8 is equal to a convective thermal resistance as it will be shown later. This happens because the contact between the two materials is not perfect due to roughness of the surface and consequentially there is a gap where the heat is exchanged by convection (R_g). Of course, there is also the resistance due to the different material in the microscopic contact areas (R_{mc}). Microscopic because the deviation from the flatness creates a non-zero roughness and so some protrusions that create the physical contact between the materials. Nevertheless, where there is a physical contact, it is possible to express the thermal resistance in a convective form. Finally, there should be also a term due to oxide films between the two materials (R_o). However, this term is usually small so it can be neglected. Therefore, the thermal resistance defined in Eq. 2.9 is given by the following sum:

$$R_{Tc} = R_g + R_{mc} + R_o \tag{2.9}$$

2.1.2 Convection

convection is a heat transfer mechanism, typical of fluids (gas or liquid), which transfers heat through its bulk motion.

The description of the convective heat transfer is more complicated than the conduction. Indeed, since we are referring to fluids, there is the necessity to take into consideration the possibility that the motion can be laminar, turbulent or transitional between the two. Moreover, the convection can be *forced* if the motion of the fluid is sustained by an external component or *natural* if the fluid is not forced by any external component. Properly taking into consideration all of these situations is not easy.

In general, there is a law that describes the convective heat transfer (Newton's law of cooling):

$$\dot{Q} = \bar{h} A \Delta T = \bar{h}(T_a - T_b) \tag{2.10}$$

where $\bar{h}$ is the convective heat transfer mean coefficient and T_a and T_b are, respectively, the surface temperature of the object (to be cooled or heated) and the fluid temperature. There is no rigorous theory of convective motion and therefore $\bar{h}$ is empirically determined being a function of the temperature, position and/or geometry of the system.

The heat exchanged can be written as:

$$Q = \frac{\Delta T}{1/\left(A\bar{h}\right)} = \frac{\Delta T}{R_{conv}} \tag{2.11}$$

where, also in this case, the convective thermal resistance has been introduced.

2.1.3 Radiation

Radiation heat transfer can play a significant role in cryogenics. Since the specific heat and cooling power involved in sub-K systems is relatively low, any source of unwanted heat source must be minimized. In laboratory conditions, the ambient acts as a blackbody source at $T = 300$ K completely surrounding the cryogenic equipment. Vacuum prevents gas conduction (and convection) while low thermal conductivity materials are used to connect stages at different temperatures. In order to minimize the effect of radiation, cryogenic systems use radiation shields.

A body at temperature T, which is a perfect radiator and a perfect absorber, emits radiation according to *Planck's law*:

$$u_\lambda(T) = \frac{2\pi hc^2}{\lambda^5 \left[\exp\left(hc/(k_B T\lambda)\right) - 1\right]} \tag{2.12}$$

where λ is the wavelength, h, c and k_B are Planck's constant, the speed of light and Boltzmann's constant, respectively. The form of this law is shown in Fig. 2.1. $u_\lambda(T)$ is measured in $W/(m^3 sr)$, so in order to compute the heat flux, we need to integrate over the wavelengths and the solid angle:

$$e(T) = \int_\Omega \int_\lambda u_\lambda(T) d\lambda d\Omega \tag{2.13}$$

If we consider that the emission is over a semi-sphere, the integration over all wavelengths gives:

$$q(T) = \sigma T^4 \tag{2.14}$$

where σ is the Stefan-Boltzmann constant. Eqs. 2.12 and 2.14 are valid only in the case of a perfect radiator. In general, for real bodies, we need to consider their emissivity, i.e., the ratio of the radiation emitted by the real body over the radiation emitted by a black-body. A body with emissivity $\epsilon < 1$ is called a *grey-body*. We have:

$$q(T) = \epsilon\sigma T^4 \tag{2.15}$$

where the emissivity ϵ has a value that spans from 0 to 1. We assume that this value is constant at all wavelengths. If the value is wavelength-dependent, it needs to be included into Eq. 2.13 and integrated.

Often in cryogenic systems we have two surfaces at different temperatures facing each other. If the two surfaces are blackbodies, then there will be a net heat transfer equal to:

$$Q = \sigma A \left[(T + \Delta T)^4 - T^4\right] \tag{2.16}$$

If the emissivities of the two surfaces are ϵ_1 and ϵ_2, then the heat exchanged becomes:

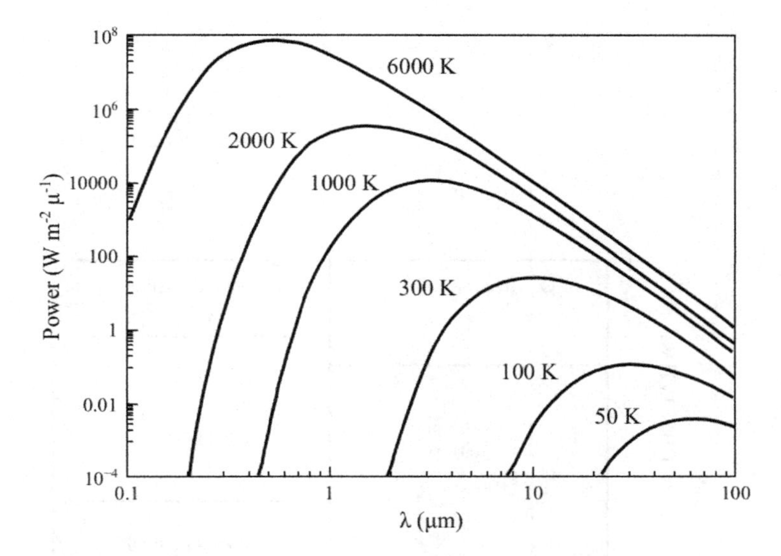

FIGURE 2.1: Power emitted by blackbodies at different temperatures

$$Q = \left(\frac{\epsilon_1 \epsilon_2}{\epsilon_1 + \epsilon_2 - \epsilon_1 \epsilon_2} \right) \sigma A \left[(T + \Delta T)^4 - T^4 \right] \quad (2.17)$$

In the case that the two surfaces are not facing directly each other, Eq. 2.17 becomes more complicated. Indeed, it is necessary to involve the *view factor* F_{ij} that takes into consideration the solid angle between the surfaces. This factor is not the same if we are considering the heat radiated by i to j or by j to i.

Adding multiple shielding can significantly reduce the heat transfer (multi-layer insulation sometimes referred to as super-insulation). If the layers have all the same emissivity ϵ, then we have:

$$Q = \left(\frac{\epsilon}{(n+1)(2-\epsilon)} \right) \sigma A \left[(T + \Delta T)^4 - T^4 \right] \quad (2.18)$$

It is very difficult to model the emissivity of the various metals used at cryogenic temperatures. There is a huge variation between metals, surface treatment, alloy composition.

In Fig. 2.2 we report a very indicative range of emissivities. Note that, as a general rule, emissivities of metal decrease with decreasing temperatures. See [64] for recent measurements of aluminium, copper, zinc, brass and stainless steel.

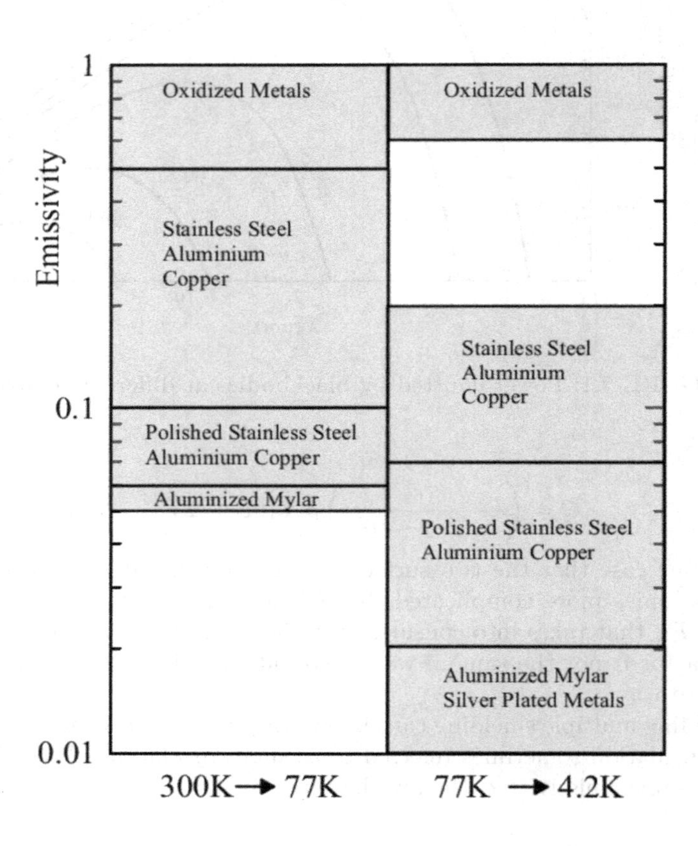

FIGURE 2.2: Emissivity of various metals used at cryogenic temperatures averaged in the temperature interval 300 to 77 K and 77 to 4.2 K. Note the large variation of values reflecting uncertainties in surface treatment and alloy composition.

2.1.4 Kapitza resistance

The thermal impedance occurring at the interface between two different mediums is called Kapitza resistance R_K. In cryogenic apparatus, especially at temperatures well below 1 K, the Kapitza resistance needs to be taken into account in order to have a proper designed system. Kapitza resistance plays an important role in all the situations where liquid helium is in contact with a metal and we require good thermal contact. For an excellent treatment of Kapitza resistance, see Ref. [80]. Here we give just a brief description.

An example of the importance of Kapitza resistance is a sub-K counterflow heat exchanger where we want to transfer effectively heat from the liquids flowing in the two opposite directions[1]. In such device, we want to have a net flow of heat from the hotter fluid to the colder fluid. This is achieved by having the phonons in the hotter fluid first interact with the metal surface (liquid/metal) of the containing tube. Then the phonons have to pass the second interface metal/liquid to finally be absorbed by the colder fluid. We are not considering other carriers of heat like, for example, electrons.

Where phonons are the main carriers of thermal energy, Kapitza resistance is due to the change in the propagation speed in the two different mediums in complete analogy to the reflection/refraction phenomena of light propagation at the interface between two mediums with different dielectric constant.

If we again limit our description to phonons as main carriers of thermal excitations, there are two mechanisms that are invoked to explain the boundary thermal resistance: Acoustic Mismatch Model (AMM) and the Diffuse Mismatch Model (DMM). See Fig. 2.3 for a simple picture of the two mechanisms. The thermal resistance is due to the crossing of phonons across the surface (in the case of Fig. 2.3, the crossing from liquid to solid). At low temperatures for liquid helium and copper as liquid and solid, the two phonon polarizations (longitudinal and transverse) behave similarly and therefore we will not separate the two cases. The geometry of the reflection/refraction at the surface is described by the equivalent of Snell's law for phonons:

$$\frac{\sin\theta_i}{\sin\theta_r} = \frac{v_i}{v_r} \tag{2.19}$$

where $\theta_{i,r}$ are, respectively, the incidence and refraction angles of the phonons and $v_{i,r}$ are the propagation speed of the phonons, respectively, in the liquid and solid mediums. We have total reflection (no thermal conduction) when $\sin\theta_r = 90°$. In the case of liquid ^{4}He $v_{liqHe} = 238\ ms^{-1}$ (for liquid ^{3}He $v = 183\ ms^{-1}$) and for copper $v_{Cu} \simeq 5{,}000\ ms^{-1}$. Using Eq. 2.19, we see that the maximum incidence angle capable of transferring phonons from liquid helium to copper is only $\sim 3°$. Therefore, the majority of phonons, i.e., those arriving at the interface liquid helium/copper with angles $\theta > 3°$ will be subject to total internal reflection and they will not be transferring heat

[1]Other heat exchangers design can have the liquids flowing in the same directions. The argument stays exactly the same.

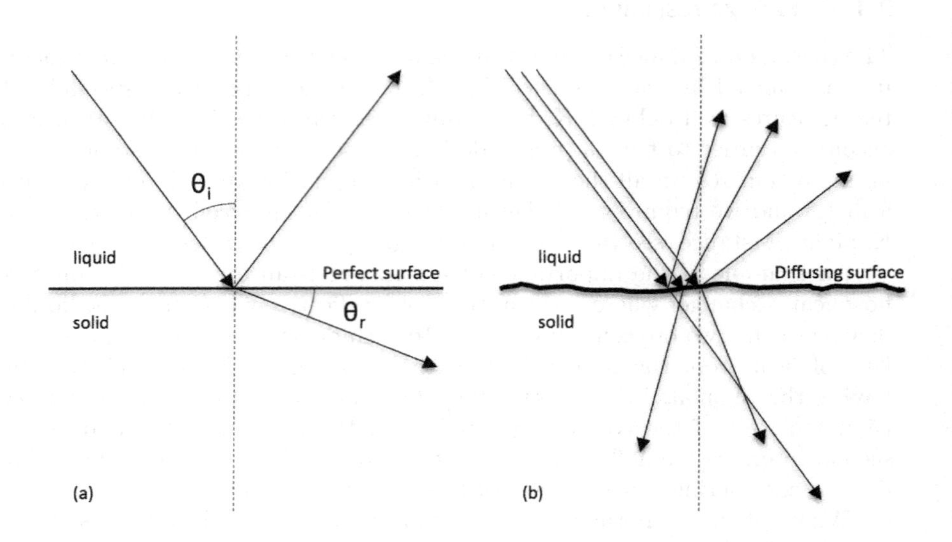

FIGURE 2.3: (a) Acoustic Mismatch Model (AMM); (b) Diffuse Mismatch Model (DMM)

from the liquid helium to the copper. The mechanism just described is termed AMM and is visually depicted in Fig. 2.3 (a). It is clear that this mechanism can severely affect the ability of heat transfer between different mediums at very low temperatures.

The expression of the power transferred between different mediums resembles Stefan's law for the blackbody radiation:

$$\dot{Q} = \frac{A}{4R_K}(T_{hot}^4 - T_{cold}^4) \tag{2.20}$$

where A is the common surface and R_K is the Kapitza resistance. The thermal boundary resistance at low temperatures is defined as the ratio between the temperature discontinuity at an interface of area A divided by the thermal power flowing through the same area A. Using Eq. 2.20 we have that the thermal boundary resistance R is

$$R = \frac{\Delta T}{\dot{Q}} = \frac{R_K}{AT^3} \tag{2.21}$$

showing that the thermal boundary resistance is proportional to T^{-3}. It is important to point out that the AMM is always predicting higher Kapitza resistances than the measured ones. Several experiments [80] have shown that one possible explanation for the overestimation of the Kapitza resistance can be traced to the quantum nature of the liquids involved, although some other

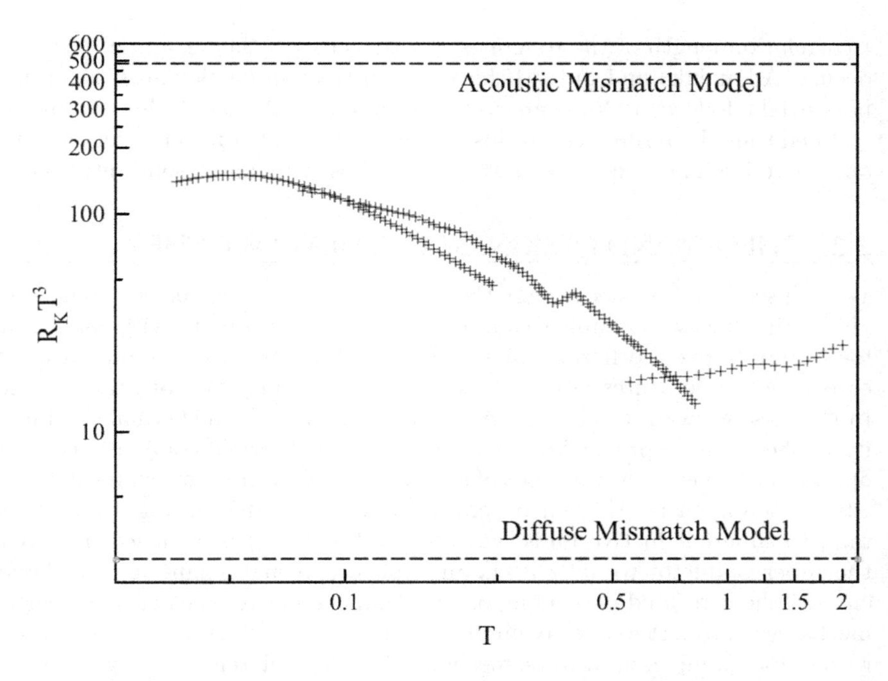

FIGURE 2.4: Measured Kapitza resistance between liquid helium and copper (data from [80] and references therein)

experiments using extremely polished surfaces and non-quantum liquids have shown that the thermal resistance was very close to the resistance predicted by the AMM.

The basic idea of DMM theory consists in assuming that all phonons impinging a surface scatter once and then are transmitted in the second medium with a probability proportional to the density of states of the phonons in the respective mediums.

Fig. 2.4 shows the measured Kapitza resistance, multiplied by T^3, for temperatures between 2 K and 0.1 K. It is clear that AMM and DMM, at best, can give respectively an upper and lower limit to the Kapitza resistance. AMM and DMM predict for liquid helium/copper interface Kapitza resistance the values of R_{KAMM} = 500 K^4cm^2W and R_{KDMM} = 2.8 K^4cm^2W. It is also clear that all the measurements are showing significant deviation from the T^3 dependence.

As of today, the discrepancy between predicted and measured Kapitza resistance is not yet fully explained. Several theoretical attempts have been made to explain the discrepancy. We want to cite the works of Adamenko and Fuks [1] and Ramiere et al. [68] where a promising mechanism is described. When the wavelength of the thermal phonons (λ) is of the order of the characteristic dimension of the roughness (σ, root mean square roughness height and

ℓ correlation length of the roughness) of the surface, then a spatial resonance occurs. Adamenko and Fuks [1] have shown that, in particular, if the roughness height follows a Gaussian distribution and $(\ell/\lambda) \approx 0.3$ phonons become trapped thus the resonance. In this condition the heat transfer is very effective and there is maximum transfer of energy across the liquid/solid interface.

2.2 THEORY AND DESIGN OF GAS HEAT SWITCHES

We will see in the next chapters that it is extremely convenient to have a device with which we are able to modulate the thermal conductivity connecting, for example, two isothermal plates. Pre-cooling from room temperature to cryogenic temperatures of cryostats is a typical application of heat switches. In this case we want to have a switch with extremely high thermal conductivity in the phase of pre-cooling and extremely low thermal conductivity in the operational phase. Several types of switches are used, from mechanical devices where the on/off is achieved by physically opening and closing a metal contact, to superconductive films/wires where the on/off is achieved by driving the superconductor to normal via an applied external magnetic field. Different switches are used depending on the temperature regimes or when certain mechanical properties are required: switches able to survive the launch of a rocket, for example, need to be mechanically very different from switches used in systems used on the ground. Some switches, for example those relying on convective motion of fluids caused by gravity, will not operate at all in space.

2.2.1 Thermal conduction of helium gas

In the temperature range of sorption coolers, temperatures are always comprised between ~ 50 K and ~ 1 K and the preferred switches are those where the thermal conducting medium is a gas (either ^{4}He or ^{3}He).

The dimensionless Knudsen number plays an important role in the theory of gas heat switches. The Knudsen number is defined by:

$$k_n = \frac{\ell}{d} \tag{2.22}$$

where ℓ is the mean free path and d is a characteristic dimension of the system. d could be the diameter of a tube where the gas is flowing or the distance between two plates or the diameter of an orifice. The Knudsen number gives a parameter to discern between different regimes. If $k_n < 0.01$, it means that the free mean path is much smaller than the typical dimensions and collisions between molecules are much more frequent than collisions with the system. In this condition, the gas can be modeled as a continuum fluid with no regards to the discrete nature of the molecules. In this regime, the thermal conductivity is highest. This regime is typical of low vacuum and, sometimes, is referred to as *laminar* or *viscous* flow. On the other hand, a Knudsen number $k_n > 1$ requires statistical mechanics for a proper theory and the discrete

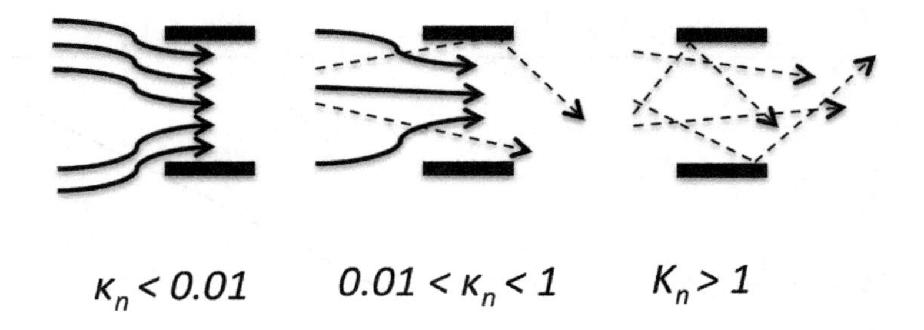

FIGURE 2.5: Laminar, Knudsen and viscous regimes

nature of the gas plays an important role. Collisions among molecules are much less frequent than collisions with the confining walls and statistical mechanics is needed for a full description. This regime is referred to as *molecular* flow typical of high/ultra-high vacuum. The intermediate region, $0.01 < k_n < 1$ is sometime termed as *Knudsen flow* and is typical of medium vacuum. These flow regimes are shown in Fig. 2.5.

In Section 1.3.3 we already discussed the thermal properties of an ideal gas where we have derived the thermal conductivity in terms of the mean free path, specific heat, average velocity, etc. When the pressure is not too high and assuming the gas to be perfect, then the heat capacity is simply $C_V = 3R/2$ (monatomic gas) and the thermal conductivity simplifies to:

$$\kappa = \epsilon \eta C_V \tag{2.23}$$

where η is the viscosity and ϵ takes into account the statistical distribution of velocities of the gas. Eq. 2.23 is useful to obtain the thermal conductivity from measurements of viscosity. For a monatomic gas $\epsilon = \frac{(9\gamma-5)}{4} = 2.5$. Assuming a Boltzmann distribution of velocities, we can make explicit the temperature dependence of the thermal conductivity:

$$\kappa(T) = \frac{\epsilon}{d^2}\sqrt{\frac{k_B^3 T}{m\pi^3}} \tag{2.24}$$

where d is the classical diameter of the molecule assumed as a rigid spherical body. Eq. 2.24 shows (again) that the thermal conductivity of a gas does not depend on the pressure. In a regime where the inter-molecular forces are important, for example in laminar flow, the viscosity itself depends on the temperature. If we consider the molecules as rigid bodies subject to attractive forces, Eq. 2.24 becomes:

$$\kappa_{lamin}(T) = \kappa_0 \frac{(C_S + T_0)}{(1 + \frac{C_S}{T})}\sqrt{\frac{T}{T_0^3}} \tag{2.25}$$

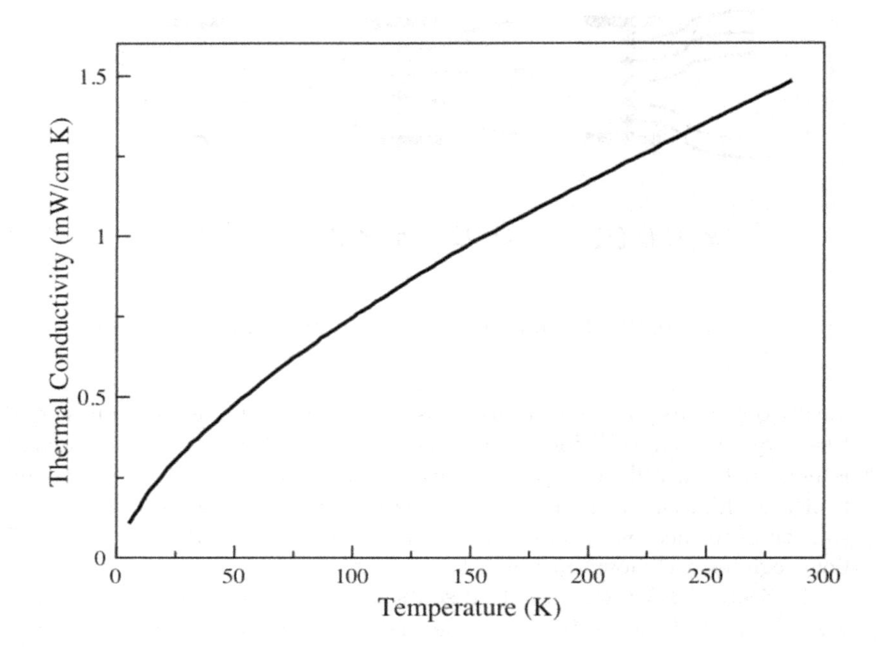

FIGURE 2.6: Thermal conductivity of ^{3}He gas

where C_S is the *Sutherland* coefficient which needs to be determined experimentally [23]. For ^{3}He gas [84] we have $C_S = 7.95$, although Eq. 2.25 is not valid for $T < 10$ K where quantum mechanical effects become important.

In Fig. 2.6 the thermal conductivity of ^{3}He gas is reported. The data provided are useful when designing thermal heat switches based on ^{3}He gas in the laminar flow regime.

For completeness we need to consider also the thermal conductivity of ^{3}He gas in the molecular regime. We give a formula for the conductance G defined as:

$$G_{molec} \equiv \frac{\kappa A}{d} = \frac{A\alpha}{4}\frac{\gamma+1}{\gamma-1}\sqrt{\frac{2R}{\pi M T}}P \tag{2.26}$$

where M is the molar weight, γ is the heat capacity ratio and α [84] is an accommodation coefficient that takes into account materials, surface conditions, temperatures, etc. At very low temperatures, a mono-layer of He molecules might cover the surfaces containing the gas: in this case, $\alpha = 1$.

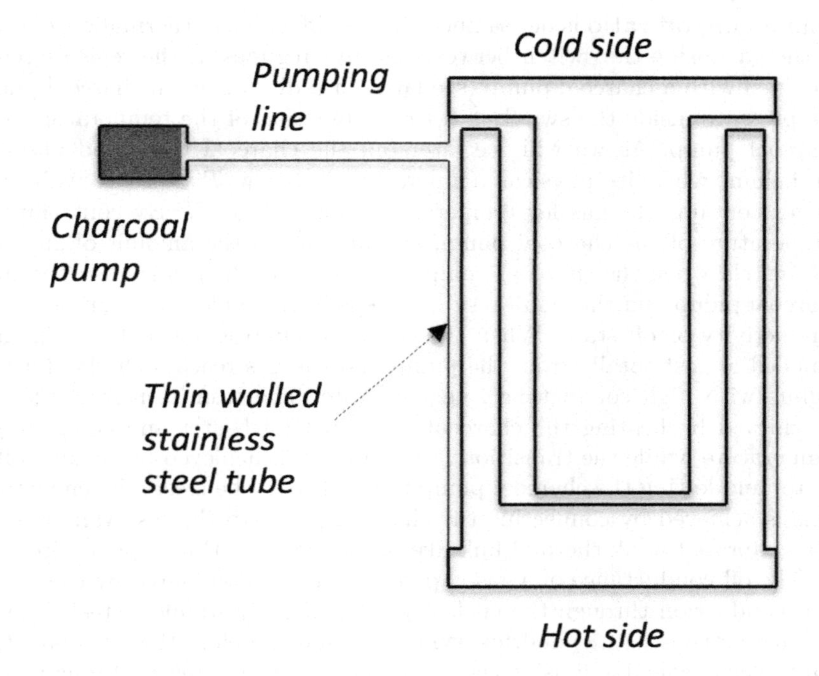

FIGURE 2.7: Sketch of a gas-gap heat switch. The weak thermal link (normally a thin copper wire) connecting the charcoal pump to the reservoir is not indicated.

2.2.2 Gas-gap heat switches

A gas-gap heat switch is a sealed device that can modulate the thermal conductivity by controlling the gas pressure between two conducting surfaces (usually copper).

In Fig. 2.7 the configuration of a typical sealed switch is shown. The two thermally isolated conductors are the *hot* conductor and the *cold* conductor. The cold conductor is usually connected to the cold reservoir (normally at $T \lesssim 4$ K), while the hot conductor is connected to the system to be either connected or isolated thermally from the reservoir. The switch is charged with helium gas and it operates between two regimes: **conducting phase** when the gas is in laminar flow and **isolating phase** when the gas is in molecular regime. We have seen that the gas conductance $G = \kappa A/d$ depends on the thermal conductivity κ, the area A where the gas is present and inversely proportional to the distance d between the hot and cold surfaces. We see therefore that to maximize the "on" conductance, we need a large area and a small gap between surfaces. These last two being geometrically fixed, the variable quantity is the thermal conductivity of the gas κ. So the important quantity **on/off ratio** is determined by the difference in thermal conductivity in the gas inside the switch between the two regimes. If the sealed circuit is provided with a charcoal pump (see later for a discussion on charcoal pumps) the pressure inside the switch is a strong function of the temperature of the charcoal pump. As we will see later on, the charcoal is a good absorbent for helium when its physical temperature is below $T \lesssim 10$ K while it no longer contains the gas for temperatures above ~ 15 K. By controlling the temperature of the charcoal pump, we can control the amount of gas inside the switch: when the charcoal pump is below 10 K the gas is absorbed in the charcoal pump and the small residual gas is in the molecular regime, i.e., low conductivity or off state. When the charcoal pump is above 15 K the gas is expelled almost totally from the pump and the gas reaches the laminar flow regime with high conductance, i.e., on state. The transition from off to on is achieved by heating the charcoal pump with a heater and controlling its temperature, while the transition from on to off is achieved by turning off the heater and letting the charcoal pump thermalize at the reservoir temperature. This is achieved by connecting the charcoal pump to the reservoir by means of a calibrated weak thermal link like, for example, a thin copper wire.

The off conductance of a gas-gap heat switch is dominated by the residual heat conduction through the enclosing tube, usually stainless steel. The normal geometry is with cylindrical symmetry and therefore the off conductance can be easily calculated using the thermal conduction integral through a tube:

$$G_{OFF} = G_{molec} + \frac{A_{cs}}{L(T_h - T_c)} \int_{T_c}^{T_h} \kappa_{SS}(T)dT \tag{2.27}$$

where, G_{molec} is the conductance of the gas in the low pressure molecular regime (usually negligible), A_{cs} is the cross-section area of the stainless steel tube, L is the length of the tube, $T_{c,h}$ are respectively the temperatures of the cold and hot ends. The on/off ratio R of conductance of a gas heat switch is the ratio of the laminar/molecular conductance expressed by:

$$R = \frac{G_{ON}}{G_{OFF}} = \frac{\kappa_{lamin} A/d}{G_{OFF}} \tag{2.28}$$

The main challenge in the design and construction of a gas-gap heat switch is the realization of large areas and small gaps as well as fast switching time off/on and especially on/off. A critical design work consists in estimating the switch conductivity as a function of the charcoal pump temperature. This conductivity depends on many factors related to the geometry of the switch and, most importantly, the characteristics of the charcoal as sorbent/de-sorbent material. A full thermal model of the switch needs to consider the three regimes: molecular (off), laminar (on) and intermediate or transitional (Knudsen). The two on/off regimes have been discussed above, while the transitional regime needs to be modeled taking into account the thermal paths.

With reference to Fig. 2.8 we see that the total conductance G_{tot} of the gas-gap heat switch, when we need to consider both on and off regimes, is

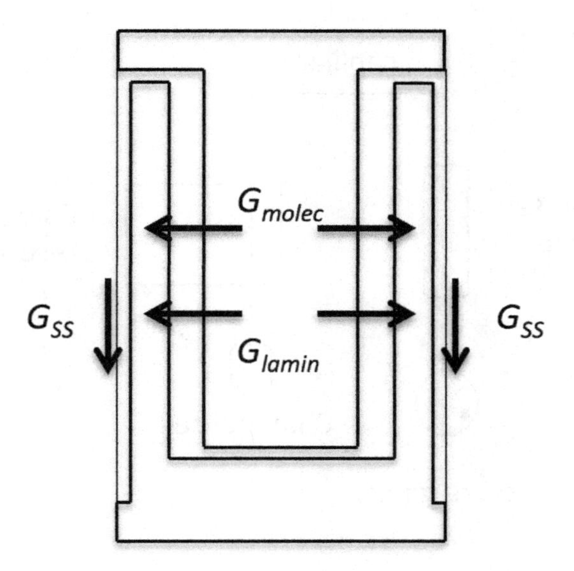

FIGURE 2.8: Thermal paths in a gas-gap heat switch

$$G_{tot} = G_{SS} + \frac{G_{molec}G_{lamin}}{G_{molec} + G_{lamin}} \tag{2.29}$$

The conductance G can be thought of as $\dot{Q}/\Delta T$, where $\dot{Q}$ is the thermal power applied to one of the two thermal surfaces of the switch and ΔT is temperature difference observed. If we consider G in Eq. 2.29 as a function of the pressure P, to evaluate the charcoal pump temperature capable of producing that pressure we need to know the adsorption isotherms of the charcoal. For a detailed discussion of charcoal pumps see Chapter 3. Here we notice that such isotherms are characteristic of the type of charcoal used. For example, a good approximation of the adsorption isotherms is

$$\ln P = [a - b\ln(q + c)](d - \frac{1}{T}) + e + \ln q + \ln 100 \tag{2.30}$$

where a, b, c, d, e are constants depending on the charcoal material used, P is the pressure and q is the amount of absorbed gas per amount of absorbent. In the case of activated coconut charcoal pellets, the coefficients in Eq. 2.30 are

$$a = 600; b = 81; c = 34; d = 0.014; e = 5.6 \tag{2.31}$$

It is difficult to predict a general form of adsorption isotherms for activated charcoal, given the variety of charcoal available and the different preparation

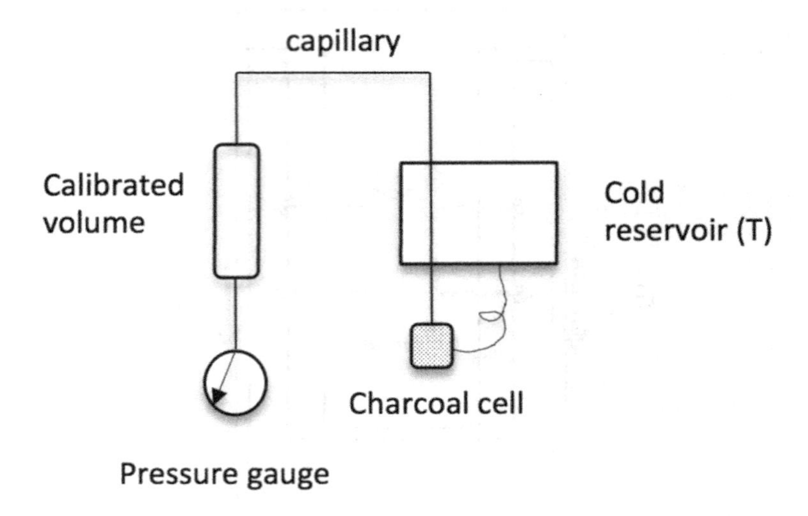

FIGURE 2.9: Schematic of an apparatus to determine absorption isotherms

processes. If accuracy of modeling is required, then the sorption isotherms should be measured experimentally [55]. A possible experimental apparatus for this measurement is shown in Fig. 2.9.

A calibrated volume at room temperature is connected through a capillary to a cell containing the known weight of charcoal to be measured. The cell is weakly linked to a cryostat containing a low temperature reservoir (a liquid helium bath or the cold head of a mechanical cooler). Once the cell has thermalized with the reservoir, heat can be applied and a finely stabilized equilibrium temperature can be reached. Helium gas can now be admitted into the calibrated volume and the pressure is recorded before and after the absorption in the cell. Several combinations of cell temperatures and gas quantities can be varied to obtain a family of isotherms. Examples of isotherms are shown in Fig. 3.3.

2.2.3 Convective heat switches

A second type of particularly effective heat switch has also been developed for use in this regime: the so-called convective heat switch [57]. In this case, the heat transfer through the switch is provided by a convective gas loop, rather than conduction through the bulk gas as described in the previous section. An example of a convective heat switch is shown in Fig. 2.10 below.

The switch essentially consists of a circuit comprising two stainless steel tubes and a copper heat exchanger at either end connected to the stages that

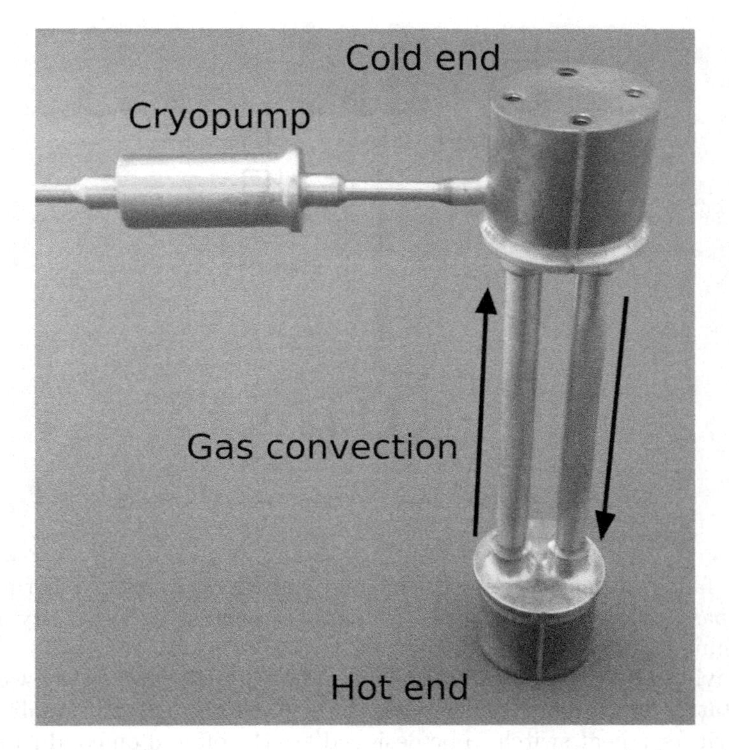

FIGURE 2.10: Convective heat switch example [57]

one wishes to couple and decouple. A charcoal cryopump (as described in detail in Chapter 3) may be used to evacuate or fill the circuit with helium gas, opening and closing the switch respectively. A cut-away schematic of the switch is shown in Fig. 2.11.

When the switch is evacuated by means of the cryopump, the only mechanism by which heat may pass through the switch is conduction through the thin-walled stainless steel tubes. Owing to the relatively low thermal conductivity of stainless steel at cryogenic temperatures (as described in Section 1.4.1) and the small cross-sectional area of the tubes, the heat flow in this condition is minimal. This is the off position of the switch, whereby the two ends of the switch (denotes "hot" and "cold" in Figs. 2.10 and 2.11 are decoupled.

In order to turn on the switch, the cryopump is heated, filling the circuit with gas. As long as the cryostat cold stages are designed so as to have the cold end positioned higher with respect to gravity than the hot end (typically achieved with copper heat straps), a convection loop (see Section 2.1.2) is established as shown in Fig. 2.11. This provides a very effective mechanism for heat transfer and hence a high conductance through the switch in the on position. The gas entering the hot end warms as it passes through the heat exchanger due to the uptake of thermal energy. It then, having warmed and

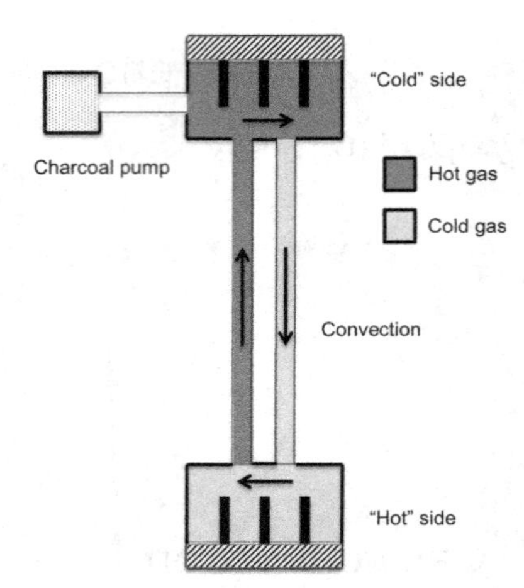

FIGURE 2.11: Convective heat switch schematic [57]

hence fallen in density, rises to the cold end. Here, it rejects thermal energy as it passes through the cold heat exchanger, increasing in density again and returning to the hot end.

Having gained a qualitative understanding of this switch, we now consider a qualitative description. As described in the previous section, the salient figures of merit for a heat switch to be designed are the off and on conductances. For the former, the only method of heat transfer when the circuit is evacuated is conduction through the stainless steel tubing and this may be calculated simply for a given tube geometry as described in Section 1.4.1.

The on conductance on the other hand is somewhat more complex; we describe here an effective model extending the work of Torii and Maris [87].

Firstly, it should be noted that the contribution from conduction through the steel remains, although ideally (in order to have a high switching ratio) this term will be dominated out by the contribution from the convective heat transfer.

In analysing the heat transfer in this case via convection, we begin by applying several fundamental laws of thermodynamics. Firstly, taking the continuity equation [94] and considering that the flow is steady, it may be seen that the quantity $\rho U A$ is constant for any "cut" through the circuit shown in Fig. 2.11, where rho is density, U is mean velocity and A is the cross-sectional area.

Next, we assume that heat transfer between the fluid and the solid components of the switch only occurs at the two heat exchangers. We can therefore consider the fluid to exist in two discrete regimes as shown in Fig. 2.11.

The next step is to consider that in the equilibrium condition, the sum of potentials around the loop is zero. In other words, the total pressure gain around the circuit has to be equal to the total pressure loss in order that the equilibrium condition is satisfied. The source of pressure rise is the buoyancy force; this occurs as we have a denser (colder) gas at the top of the switch and a less dense (warmer) gas at the bottom, with respect to gravity. The pressure gain may be found from the difference in densities as

$$\Delta P_\rho = \Delta\rho g z = (\rho_1 - \rho_2)\, g z \tag{2.32}$$

In opposition to this, we have pressure losses in the flow due to dissipative forces; essentially internal friction (viscosity) in the flow as it travels around the loop causes a drop in pressure. In the case of laminar flow, the pressure drop per length of tube L travelled is given for this geometry by the Darcy-Weisbach equation [94] as

$$\frac{\Delta P_\mu}{L} = \frac{128}{\pi} \cdot \frac{\mu Q}{D^4} \tag{2.33}$$

where μ is the dynamic viscosity of the fluid, Q is the volumetric flow rate and D is the diameter of the tube.

Considering the two flow regimes shown in Fig. 2.11, the sum of the viscous losses around the circuit is given by

$$\Delta P_{\mu,1} + \Delta P_{\mu,2} = \left(\frac{128}{\pi} \cdot \frac{\mu Q L}{D^4}\right)_1 + \left(\frac{128}{\pi} \cdot \frac{\mu Q L}{D^4}\right)_2 \tag{2.34}$$

By considering that the volumetric flow is related to mass flow (which, from continuity, we know to constant around the circuit) by

$$Q = \frac{\dot{M}}{\rho} \tag{2.35}$$

and taking the lengths of the two sections to be the same, i.e.,

$$L_1 = L_2 \tag{2.36}$$

and L is the total length of the circuit

$$L = L_1 + L_2 \tag{2.37}$$

we can re-write the sum of the viscous losses as

$$\Delta P_{\mu,1} + \Delta P_{\mu,2} = \frac{8\dot{M}L}{\pi r^4} \cdot \left(\frac{\mu_1}{\rho_1} + \frac{\mu_2}{\rho_2}\right) = \Delta P_\mu \tag{2.38}$$

Recalling that the sum of potentials around the loop is zero, we then get

$$(\rho_1 - \rho_2)\, g z = \frac{4\dot{M}L}{\pi r^4} \cdot \left(\frac{\mu_1}{\rho_1} + \frac{\mu_2}{\rho_2}\right) \tag{2.39}$$

Re-arranging now for the mass flow rate gives

$$\dot{M} = \frac{\pi r^4}{4L} \cdot \frac{(\rho_1 - \rho_2) g z}{\frac{\mu_1}{\rho_1} + \frac{\mu_2}{\rho_2}} \tag{2.40}$$

Imagine now that we draw a control volume around the heat exchanger at the hot end. From the first law of thermodynamics (see Section 1.1.2.1), it may be seen that in the steady-state condition the heat transferred into the switch at the hot end is equal to the difference in enthalpies of the outgoing and returning gas flows to the heat exchanger (similarly for the heat rejected at the cold end). This then allows us to consider that the heat through the switch $\dot{Q}$ is the product of the mass flow (constant around the circuit) and the difference in enthalpies Δh. This gives us

$$\dot{Q} = \dot{M} \cdot \Delta h = \frac{\pi r^4}{4L} \cdot \frac{(\rho_1 - \rho_2) g z}{\frac{\mu_1}{\rho_1} + \frac{\mu_2}{\rho_2}} \cdot (h_2 - h_1) \tag{2.41}$$

It may further be seen that the difference in enthalpy (neglecting pressure loss through the heat exchanger due to the short path length) is

$$h_2 - h_1 = c_v (T_2 - T_1) \tag{2.42}$$

where c_v is the specific heat capacity. It is important to note in designing this type of switch that the heat exchangers will in reality be imperfect (i.e., the exit temperature of the flow will approach but not be equal to the heat exchanger sink temperature), $T_1 \neq T_C$ and $T_2 \neq T_H$. We will return to this point shortly.

From Eq. 2.41, we can see that in order to find the power through the switch we simply need to know the densities and temperatures of the two regimes. The densities are a function of temperatures themselves, as well as the initial gas charge in the switch. It is clear from inspection that the greater the amount of gas in the switch, the greater the ON conductance of the switch. However, the practical upper limit on the charge is given by either the capacity of the cryopump (as it must be able to fully evacuate the circuit) or the mechanical limitation given by the structure in terms of the allowable charging pressure, whichever is the lower. The cryopump may be sized as described in Chapter 3 and hence in practice the switch mechanical limitation is likely to be the determining factor. This is typically found using a finite element approach, on which extensive literature is otherwise available [41] [81] [15].

Given the switch geometry and the initial charging pressure, the number of moles of gas may be calculated from the perfect gas law as described in Section 1.1.3.

By considering the two gas regimes, as shown in Fig. 2.11, the number of moles and hence the density may be calculated as a function of the two regime temperatures.

Using the model described by Eq. 2.40, it may be seen how a contour plot may be produced for the mass flow through the switch as a function of the two regime temperatures as given in Fig. 2.12 (in arbitrary units).

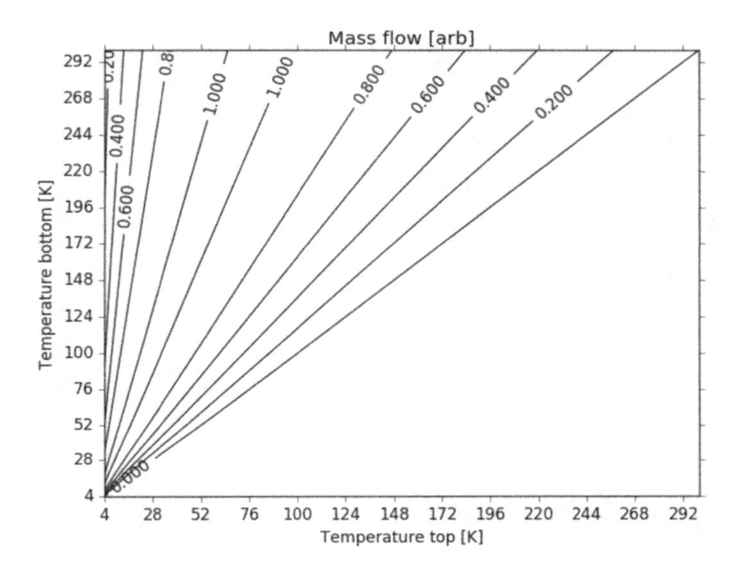

FIGURE 2.12: Mass flow in convective heat switch

Furthermore, we can see from Eq. 2.41 how this translates into the power through the switch as function of the two regime temperatures as shown in Fig. 2.13 (again, in arbitrary units).

As the exact values of the mass flow and power will depend on several factors such as the geometry, charge and heat exchanger efficiencies, the plots are given in arbitrary units as illustrative examples. The key point to note is that there exists a "sweet spot". If the temperature differential is too great, then the density in the denser region will be sufficiently high so as to result in great enough viscous losses so as to offset the increased head gain from buoyancy forces.

As was mentioned briefly earlier in this analysis, we must consider that the heat exchangers are imperfect and hence their temperatures are not equal to the flow exit temperatures. In order that the models we have developed have some predictive power (i.e., we are able to calculate the power through the switch as a function of the end temperatures for a given design), we must find a relationship between the two. In other words, we must be able to characterise the efficiency of the heat exchangers.

Helpfully, the heat exhanger geometries in this case are identical and as the mass flow through both is the same, we make the assumption that the efficiency of each is also the same. We can then consider a model where the inefficiency of both exchangers is parameterized by a single value f as

$$T_1 = T_C + (f \cdot (T_H - T_C)) \tag{2.43}$$

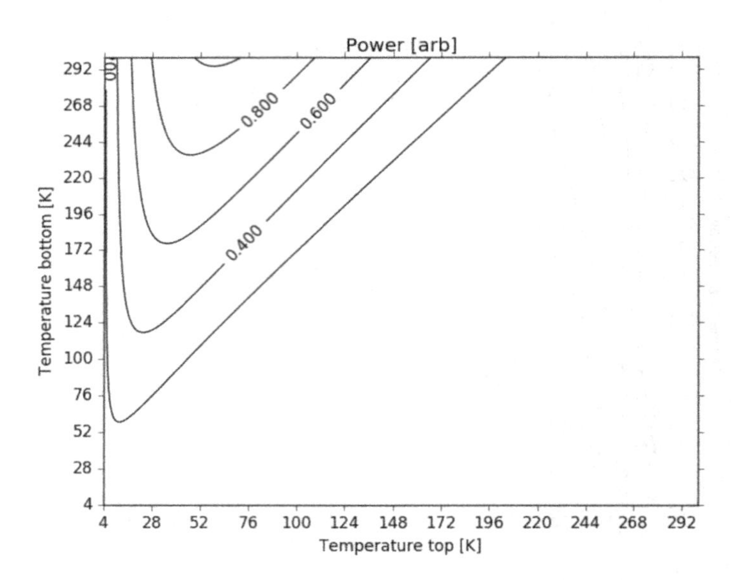

FIGURE 2.13: Heat flow in convective heat switch

$$T_2 = T_C + ((1 - f) \cdot (T_H - T_C)) \tag{2.44}$$

Unfortunately, modeling of f from first principles is exceptionally difficult owing to the complex geometries used to maxmise the surface area and the flow conditions in this part of the circuit. In our experience, a heat exchanger efficiency on the order of 5% has been found to be a reasonable value for the purposes of switch design, although we would of course encourage the readers to validate their own heat exchanger designs experimentally.

FURTHER READING

Barron, R.F. and Nellis, G.F. (2017). *Cryogenic Heat Transfer.* CRC Press.

Pobell, F. (2007). *Matter and Methods at Low Temperatures.* Springer.

CHAPTER 3

Sorption Cryopumps

IN those cryogenic applications, such as large liquid bath cryostats used to cool superconducting magnets and radio frequency cavities found in modern particle accelerators, pumping is done on the working fluid by mechanical pumps. However, for the much lower heat loads (hence lower throughputs), lower temperatures (hence lower pressures) and smaller mechanical size of coolers for astrophysical applications, pumping provided by adsorption onto a large surface is the preferred option.

Charcoal pumps using helium molecules are one of the important components in a sub-K sorption cooler. The charcoal pump is usually made of a canister filled with charcoal pellets (or other suitable material) in good thermal contact with a thermal link which, in turn, is connected to a low-temperature reservoir (~ 4 K) through either a weak thermal link or a heat switch. In the next chapter we will discuss the details of how to design and build such a charcoal pump (see Fig. 4.4 for a schematic view, and Fig. 4.5 for a picture of the inside of a charcoal pump). Depending on the temperature of the charcoal, it can act as a high efficiency pump (at low temperatures) or it can desorb the gas previously captured (at higher temperatures).

Simply by heating or cooling the charcoal inside the pump we can switch between the pumping or desorbing states. Since both the heating and the cooling are achieved by means of electrically operated heaters[1], the operation can be easily computer controlled with no vibrations and no valves to be operated resulting in a very practical and compact system.

In order to design such a pump, a good understanding of the sorption process (in particular physical adsorption) and behaviour of sorbent materials is required.

[1]For the convenience of the reader, we repeat here briefly the operations of a gas heat-switch already discussed in Chapter 2. We have seen that by heating the small charcoal pump of a gas-gap or convection switch, we turn the switch to a high thermal conductivity state. Low conductivity state is achieved by removing electrical power and allowing the small charcoal pump to cool to the reservoir temperature by means of a weak thermal link. When the charcoal is cold, exchange gas inside the switch is absorbed and the switch is turned off.

3.1 PRINCIPLES OF PHYSISORPTION

As a broad definition, adsorption is simply the adhesion of atoms, ions, or molecules to a given surface. A given adsorption process may be considered physical (physisorption) or chemical (chemisorption), depending on the forces governing the interaction.

Physisorption describes a mechanism of adsorption in which the forces involved are intermolecular (the sum of which are the van der Waals forces). This is in contrast with chemisorption, in which the forces involved are valence forces (of the same kind as those operating in the formation of chemical compounds).

The binding energies for chemical adsorption are much higher (>0.5 eV per species adsorbed) and hence for reasons that are discussed below, only physisorption will be considered for cryogenic sorption pump design.

A theoretical description of the physisorption mechanism allows some qualitative understanding.

3.1.1 Van der Waals forces and the Lennard-Jones potential

The behaviour of an idealised gas may be described by the classic equation of state (3.1).

$$PV = nRT \tag{3.1}$$

However, as this idealisation neglects intermolecular forces (it may be derived simply from kinetic theory), it is not suitable for modeling the physisorption process.

The sum of all the forces that are not related to an electrostatic interaction or a covalent bond are grouped together into a unique category: the *van der Waals Force*. This force includes three major contributors: the *London Dispersion Force*, the *Keesom Interaction* and *Debye Force*. The London Dispersion Force is due to the interaction between two non-permanent dipoles. The first non-permanent dipole is generated by the movement of the electrons that can create an instantaneous dipole. Consequentially, another atom that passes close to the first will be polarized by the electric field created. These two atoms are hence coupled together and as a result, any change that is experienced by one electron cloud will affect the distribution of the electrons in the second cloud. The Keesom Interaction is due to the interaction between the two permanent dipoles. Finally, the Debye Force is due to the interaction between a permanent dipole and an induced one. The induced dipole is created in the same way described for the London Force, the only difference being that the first dipole is in this case permanent.

These forces are attractive and relatively long-range. However, it is also necessary to consider the effects of the repulsive forces that become dominant when two atoms or molecules are closer together. These forces are mainly due to the nuclear and electronic repulsions. Eq. 3.1 may be modifed to account for the presence of these forces. Indeed, it is possible to consider the atoms as spheres, having a non-zero volume. In this case, the volume available for the movement of the atoms is reduced. Therefore, the volume occupied by the gas is not V, but it is $V - nb$ where nb is the volume occupied by all the atoms. This term is due to the presence of the repulsive forces that do not allow particles to occupy the same physical space. Eq. 3.1 now becomes:

$$P(V - nb) = nRT \tag{3.2}$$

Moreover, it is necessary to consider that, due to the presence of the attractive forces, the pressure is reduced. Indeed, their net effect is to reduce the frequency and the force of the collisions between the particles with each other and the wall of the container occupied by the gas. This effect can be quantified as $-a(n/V)^2$, where n/V is the molar concentration that is proportional to the presence of the attractive interactions. This quantity is squared because the attractive forces act both on the frequency and the force of the collisions. Given these considerations, it is possible to write the *van der Waals equation* as:

$$P = \frac{nRT}{V - nb} - a\left(\frac{n}{V}\right)^2 \tag{3.3}$$

This equation takes into consideration the presence of both repulsive and attractive forces; the constants a and b are specific to a given gas species.

The van der Waals equation of state takes into consideration the presence of intermolecular forces, but it does not quantify them. The first attempt to do that was made by John Lennard-Jones in 1924 [42] when he introduced a potential that accounts for both the repulsive and attractive forces[2]. This potential is described by Eq. 3.4 and it may be seen that it contains two components: an attractive term proportional to r^{-6} (long range) and a repulsive term proportional to r^{-12} (short range).

$$U(r) = -4\epsilon\left(\frac{\sigma}{r}\right)^6 + \left(\frac{\sigma}{r}\right)^{12} \tag{3.4}$$

where ϵ is the depth of the potential well and σ is the distance where the potential is zero. A representative plot of this potential is shown in Fig. 3.1. As evident from the plot, after $r/\sigma \simeq 2$ the potential is flattening to zero, this means that intra-molecular attraction is almost zero and the two particles are not bounded by any force. The increasing of the potential at low value of r/σ is due by the Pauli repulsion. This effect visible in the plot is clearly derived from the r^{-12} term in Eq. 3.4. The most important characteristics of the potential is the presence of the well located at $r = 2^{1/6}\sigma$. This particular position is the intra-molecular distance when the two molecules (or atoms) are bound together. The correspondent energy is binding energy.

[2]We have already briefly discussed the Lennard-Jones potential for helium in Chapter 1.

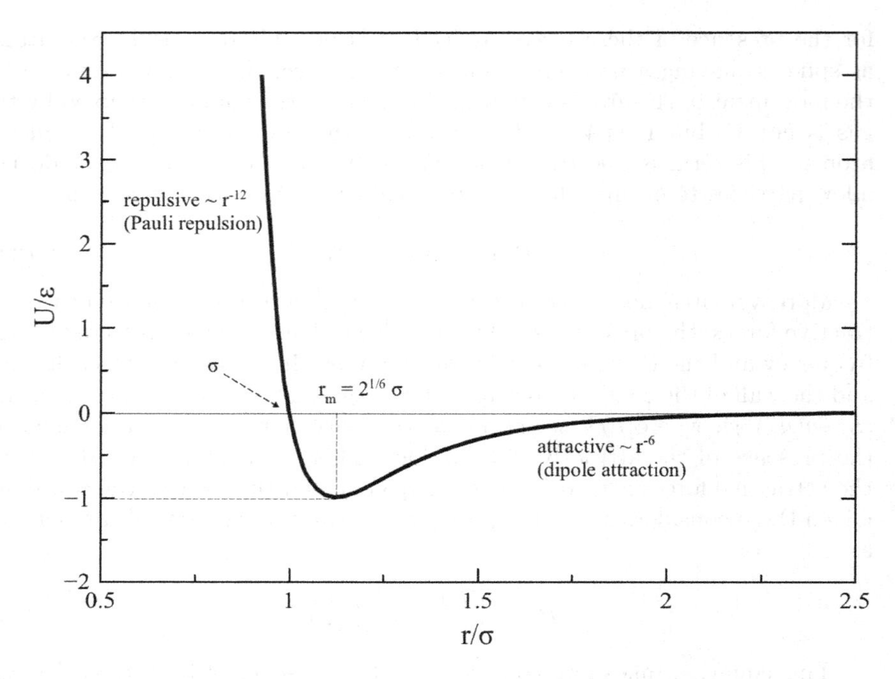

FIGURE 3.1: Graph of general Lennard-Jones potential

3.1.2 Physisorption theory

A particular feature of the Lennard-Jones potential is given by the well in which particles of a sufficiently low kinetic energy may become trapped. It may therefore be seen how gas molecules may be physisorbed onto the surface of a suitably cold sorbent.

The adsorption of gases onto the surface of different materials had been observed before the publication of the Lennard-Jones theory. Indeed, in 1918, Irving Langmuir had attempted to explain this phenomenon. In order to do so, he considered the sorbent surface as planar and composed of different elementary spaces that can adsorb only one molecule each. Consequentially, a layer of adsorbed gas one molecule thick is formed over the sorbent and so, when the layer is "full", no other molecules may be adsorbed. The rate of adsorption by the material is $\alpha\mu$, where μ is the number of molecules, in moles, that hit the sorbent surface per cm^2 and second and α is the fraction of the molecules that remain trapped on the surface. In Langmuir theory, only the empty cells on the sorbent surface are considered for the adsorption process and so the total rate of adsorption is given by $\alpha\theta_0\mu$ where θ_0 is the fraction of free cell on the sorbent surface, so available for the adsorption process.

Until now, we have considered only the adsorption process, but we need to take into consideration also the inverse process: desorption. The rate of desorbed molecules per cm^2 and second is given by $\nu\theta_i$ where ν is the rate at which molecules would be desorbed if every cell is filled and θ_1 is the fraction of the cell occupied on the sorbent surface. At the equilibrium these two rates should be equal, so:

$$\alpha\theta_0\mu = \nu\theta_1 \tag{3.5}$$

Moreover, we have the condition that $\theta_0 + \theta_1 = 1$, so:

$$\theta_1 = \frac{\alpha\mu}{\nu + \alpha\mu} = \frac{\sigma_1\mu}{1 + \sigma_1\mu} \tag{3.6}$$

where it has been introduced $\sigma_1 = \alpha/\nu$ that is called the *relative life*. The fraction of occupied cell can be written also as:

$$\theta_1 = \eta\frac{N_A}{N_0} \tag{3.7}$$

where N_A is the Avogadro number, N_0 is the number of cells per cm^2 on the surface and η is the number of molecules, in moles, adsorbed per cm^2 of sorbent surface. Combining Eq. 3.7 and Eq. 3.6, it is possible to write the so-called *Langmuir equation*:

$$\eta\frac{N_A}{N_0} = \frac{\alpha\mu}{\nu + \alpha\mu} = \frac{\sigma_1\mu}{1 + \sigma_1\mu} \tag{3.8}$$

Until now, the sorbent surface has been considered as an ensemble of elementary cells with the same properties. However, this condition is not always true. More realistically we can have that the surface is composed of several kinds of cells, so:

$$\sum_{i=1}^{n} \beta_i = 1 \tag{3.9}$$

where β_i represents the fraction for every kind of cell. Each kind of cell acts independently, so it is possible to superimpose a set of Langmuir equation for each fraction β_i:

$$\eta\frac{N_A}{N_0} = \sum_{i=1}^{n} \frac{\beta_i\sigma_i\mu}{1 + \beta_i\mu} \tag{3.10}$$

The Langmuir equation can be written in a different way making these assumptions: (a) the number of molecules that hit the sorbent is proportional to P, which is the pressure of the gas, so $\mu \propto P$, and (b) σ_1 quantifies the efficiency of the adsorption process and can be expressed as a solution of an Arrhenius equation, so $\sigma_1 \propto \exp(-\Delta E_1/RT)$ where ΔE_1 is the difference in energy between the adsorbed state and the free one.

Therefore, Eq. 3.8 becomes:

$$\eta \frac{N_A}{N_0} = \frac{k_1 P e^{-\Delta E_1/RT}}{1 + k_1 P e^{-\Delta E_1/RT}} \tag{3.11}$$

where k_1 is a constant that takes into consideration both proportionality factors for μ and σ_1.

With the previous assumptions, it is also possible to rewrite Eq. 3.5 for the equilibrium:

$$a_1 P \theta_0 = b_1 \theta_1 e^{-\Delta E_1/RT} \tag{3.12}$$

where a_1 and b_1 are proportionality constants between P and μ and between σ_1 an $\exp(-\Delta E/RT)$[3], respectively.

The Langumir theory describes only the single layer adsorption, so when all the sorbent surface is filled, it is not possible that any other molecules can be adsorbed. However, it is necessary to consider also the possibility that more than one layer of adsorbed molecules is created. Indeed, the van der Waals force does not act only between the molecules and the sorbent surface, but also between the molecules themselves. Moreover, it is not possible to exclude completely that an adsorbed molecule *shields* the interaction between the sorbent and the other free molecules. An extension of the Langumir theory was proposed in 1938 [11] by Brunauer, Emmett and Teller. Their theory extends the equilibrium Eq. 3.12, considering that now the first layer can be populated by the condensation on the bare surface from the second layer or the free space. Instead, the evaporation towards the free space or the second layer contributes to a depopulation of the first layer. At the equilibrium, the condensation rate and the evaporation rate should be equal, so:

$$a_1 P \theta_0 + b_2 \theta_2 e^{-\Delta E_2/RT} = b_1 \theta_1 e^{-\Delta E_1/RT} + a_2 P \theta_1 \tag{3.13}$$

where a_2 and b_2 have similar meaning of a_1 and b_1 and ΔE_2 is the difference in energy between the first and the second layers. Here, θ_i represents the cells available for creating the $i+1$ layer, so it is the ratio between the number of cells with i adsorbed molecules and the total number of cells. Combining Eqs. 3.12 and 3.13, it is possible to find the following relation:

$$a_2 P \theta_1 = b_2 \theta_2 e^{-\Delta E_2/RT} \tag{3.14}$$

Now, it is possible to extrapolate this result for every layer, so:

$$a_i P \theta_{i-1} = b_i \theta_i e^{-\Delta E_i/RT} \tag{3.15}$$

The total available area for adsorption is given by:

$$A = \sum_{i=0}^{\infty} \theta_i \tag{3.16}$$

[3]It is possible to have σ_1 in Eq. 3.5 simply dividing all the equations by α.

while the total volume is:

$$v = v_0 \sum_{i=0}^{\infty} i\theta_i \tag{3.17}$$

where v_0 is defined as the volume occupied by the first layer only (completely filled) in one cm^2. Consequentially, the total volume of the first layer is Av_0. Now, it is possible to compute the ratio between the total volume of gas adsorbed and the volume of the mono-layer, which is equal to

$$\frac{v}{v_m} = \frac{\sum_{i=0}^{\infty} i\theta_i}{\sum_{i=0}^{\infty} \theta_i} \tag{3.18}$$

Now, it is possible to simplify the problem considering that the amount of energy requested for moving between difference layers is always the same, so $\Delta E_2 = \Delta E_3 = \cdots = \Delta E_L$ where ΔE_L is the liquefaction energy. Moreover, it is possible to consider that $b_2/a_2 = b_3/a_3 = \cdots = g$. These two simplifications imply that the evaporation-condensation properties of the i-th layer with $i > 1$ are equal to those of the liquid state.

With these assumptions, it is also possible to re-write the term θ_i as function of θ_0:

$$\theta_1 = y\theta_0 \quad y = a_1/b_1 e^{\Delta E_1/RT} \tag{3.19a}$$

$$\theta_2 = x\theta_1 \quad x = p/g e^{\Delta E_L/RT} \tag{3.19b}$$

$$\theta_3 = x^2\theta_1 \tag{3.19c}$$

$$\theta_i = x^{i-1}\theta_1 = yx^{i-1}\theta_0 = cx^i\theta_0 \tag{3.19d}$$

where in the last equation, the factor $c = y/x$ has been introduced. So, inserting Eq. 3.19 in 3.18, the ratio between v and v_m becomes

$$\frac{v}{v_m} = \frac{c\theta_0 \sum_{i=1}^{\infty} ix^i}{\theta_0 \left(1 + c\sum_{i=1}^{\infty} x^i\right)} \tag{3.20}$$

The sum in the denominator is simply a geometrical sum, so $\sum_{i=1}^{\infty} x^i = \frac{x}{1-x}$. Instead, the sum in the numerator can be computed considering $\sum_{i=1}^{\infty} ix^i = x\frac{d}{dx}\sum_{i=1}^{\infty} x^i$, so it is equal to $\frac{x}{(1-x)^2}$. With these mathematical substitutions, Eq. 3.20 becomes:

$$\frac{v}{v_m} = \frac{cx}{(1-x)(1-x+cx)} \tag{3.21}$$

In case of adsorption on a free surface, the only possibility to have an infinite number of layers is given when the $p = p_0$ where p_0 is the saturation pressure of the gas. In particular, for $v = \infty$ when $p = p_0$ the value of x has to be equal to 1. Therefore, $p_0/g e^{\Delta E_L/RT} = 1$ and $p/p_0 = x$. With this in mind, it is possible to write the *BET equation*:

$$v = \frac{cv_m p}{(p_0 - p)(1 + (c-1)p/p_0)} \tag{3.22}$$

However, the BET equation is usually written as

$$\frac{p}{v[p_0 - p]} = \frac{c-1}{v_m c}\left(\frac{p}{p_0}\right) + \frac{1}{v_m c} \tag{3.23}$$

Eq. 3.23 is hence an isotherm as shown in Fig. 3.2. Usually, to keep the plot linear, the quantity p/p_0 is used on the x-axis while $p/(v(p_0 - p)$ on the y-axis. The gradient of the resulting line is simply given by the coefficient of p/p_0, while the intercept is given by $1/(v_m c)$. Knowing the gradient and the intercept allows the computation of the BET constant c and the mono-layer volume v_m.

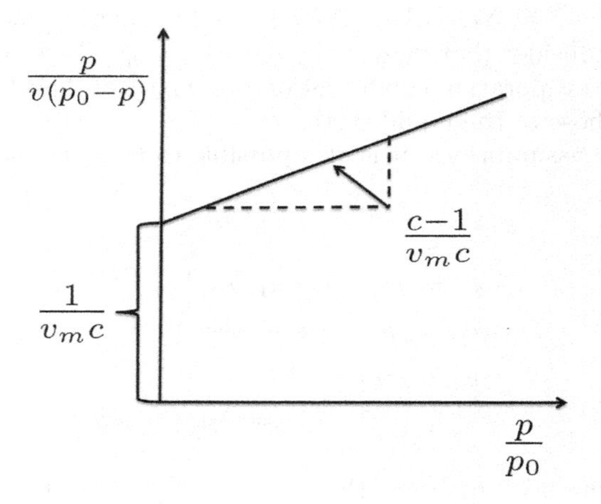

FIGURE 3.2: BET plot from Eq. 3.23

Values are hence obtained for

$$S_{total} = \frac{(v_m N s)}{V} \tag{3.24}$$

and

$$S_{BET} = \frac{S_{total}}{a} \tag{3.25}$$

where a is the mass of the solid and S_{BET} is the specific surface area. Whilst the method is not comprehensive, it allows measurements of the surface area of a material and provides a guide to sorbent material selection as explained in Section 3.2 below.

Both Langmuir and BET theories are based on the equilibrium between the adsorption and desorption of the gas on the sorbent. Instead, Dubinin proposed a theory that is based on the adsorption potential defined as[4]:

$$A = RT \log \frac{p_s}{p} \tag{3.26}$$

where p is the pressure and p_s is the saturated vapour pressure. Using this potential and considering a pore size distribution as Gaussian, he initially found:

$$W = W_0 e^{-\left(\frac{A}{\beta E_0}\right)^2} \tag{3.27}$$

where W is the current volume of adsorbed gas, W_0 is the limit volume that can be adsorbed, E_0 is called *characteristic energy* and is equal to the adsorption potential computed in a reference point and, finally, β is the ratio between A and E_0, so $\beta = A/E_0$.

However, it was found that Eq. 3.27 is valid only if the micro-pore sizes are in a very small dimension interval. In this case, Eq. 3.27 can be generalized as:

$$W = W_0 e^{-\left(\frac{A}{\beta E_0}\right)^N} \tag{3.28}$$

where N is a constant. This formula is purely empirical, so in 1977 Stoeckli [79] proposed its rigorous correction, where the adsorbed volume is given by the contribution of different micro-pore size groups. In this case, Eq. 3.27 becomes

$$W = \sum W_{0,i} \exp\left[-B_j \left(\frac{T}{\beta}\right)^2 \log_{10}^2\left(\frac{p_s}{p}\right)\right] \tag{3.29}$$

where B_j is called *structural constant* and is proportional to $(R/E_{0,j})^2$. In case of a continuous size group distribution, Eq. 3.29 becomes

$$W = \int_0^\infty f(B) \exp\left[-B \left(\frac{T}{\beta}\right)^2 \log_{10}^2\left(\frac{p_s}{p}\right)\right] dB \tag{3.30}$$

where $f(B)$ is a pore group distribution function.

3.2 SORBENT MATERIALS

All the theory presented on the adsorption of gas is necessary to understand and choose the best material for a sorption cooler cryopump. All sorbent materials are sold in the form of grains of different dimensions. Usually, the grain dimension is not important for the material selection because as shown in the previous section the adsorption is a microscopical phenomenon. However,

[4]For a proper derivation of this potential, see [10].

the dimension of grain can be important to optimize the quantity of material required. In fact, bigger grains mean that there will be more free space between each grain, so less material available for adsorption. However, smaller grains can reduce too much the free space and reduce significantly the flow of the gas. Moreover, smaller grains can also fall in the tube that transports the gas. The choice of an adsorbent material needs to be selected and be sized such that it is able to adsorb all the gas in the sorption cooler and at the same time it is able to create a good vacuum. This implies a reduction of the pressure on the liquid and consequentially a dropping in temperature. Another important characteristic of the material is related to its surface-to-volume ratio. Indeed, this needs to be elevated so that each gram of material can adsorb as much gas as possible. This quantity is also important because it means that the cryopump is not practically full with the adsorbent material. Indeed, a high surface-to-volume ratio means that that each gram of material has a lot of free space that allows the gas to flow. Therefore, not all the material needs to be close to the tube which transports the gas and it will be the gas that slowly will move to the furthermost material grains.

Several materials are available as sorbent, such as zeolite or activated charcoal. In this book, we will talk only about charcoal since it is used in almost all the sub-K cryogenic applications.

3.2.1 Charcoal

Charcoal is a material that is widely used in many adsorption applications. The great advantage of this material is that it can be produced from multiple sources, such as coal, coconut or hardwood. This is also its greatest disadvantage; in fact, it is pretty hard to estimate charcoal performances (such as pumping speed and adsorption capacity) since they are dependent on the material used to produce it. In addition, its performances are also dependent on how the material is carved and in particular how it is *activated.* Activation can be chemical if the charcoal is activated in a bath of chemicals. Otherwise, it may be activated using hot gas (exposed to it or forcing the gas to flow inside), in this case the activation is defined physically.

In general it is possible to give some general information about the charcoal that can be useful in designing a sorption cooler. In particular it is possible to notice that charcoal has a large effective surface area, more than 1000 m^2/g, regardless of the production process. This is clearly shown in Table 3.1, where several kinds of charcoal, produced by different companies (so different activation process) and from different materials, are compared.

The main characteristic of charcoal is its capacity to improve the adsorption power decreasing the temperature. Adsorbing more gas implies the capacity to create a *higher vacuum*, decreasing the charcoal temperatures. This property is at the base of every sorption cooler. In Fig. 3.3, it is possible to notice charcoal adsorption of helium at cryogenic temperatures. From this plot, it is important to notice that below 4.2 K, the adsorption power is al-

Type	Source	Mesh (mm)	Supplier	Surface (m^2/g)
PCB	Coconut	1.7-0.6	Calgon	1150-1250
BPL	Coal	1.7-0.6	Calgon	1050-1150
GAC 1240	Coal	1.7-0.4	Ceca, Inc.	1000-1100
Nuchar WV-B	Hardwood	2.0-1.7	Westvaco	1500-1700

TABLE 3.1: Surface area data for a range of charcoal samples [74]

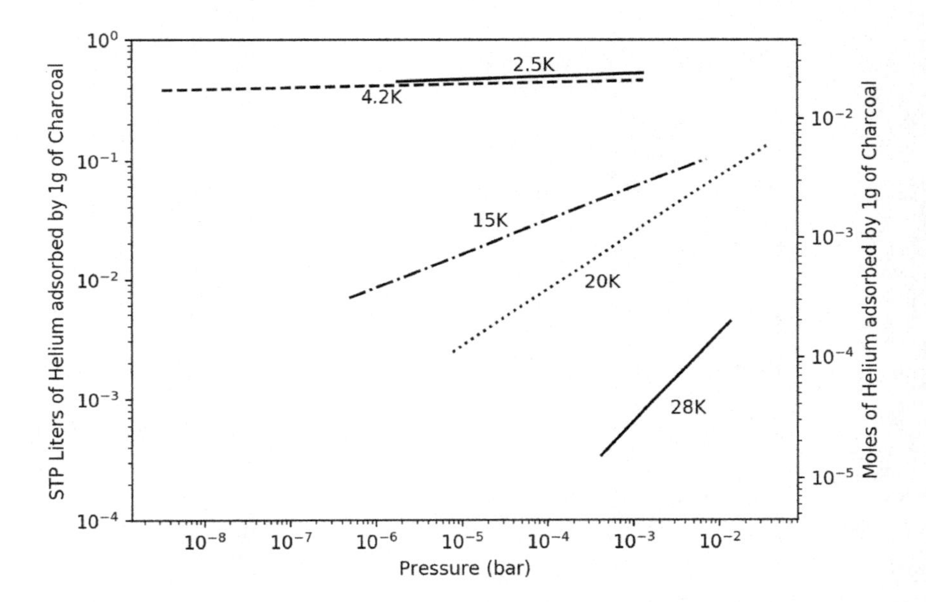

FIGURE 3.3: ^{4}He adsorption isotherms on charcoal at different temperatures (data from [66])

most constant; hence it is probably not so effective to significantly reduce the temperature of a charcoal pump further.

However, it is observed that whilst there is a general correlation between BET surface area and helium accumulation at a given temperature and pressure, there are other factors involved. As shown by Sedgley et al. [74] having the largest specific surface area does not guarantee the best performance in terms of either helium accumulation capacity or pumping speed.

FURTHER READING

Bansal, R.C. and Goyal, M. (2005). *Activated Carbon Adsorption.* CRC Press.

II

Applications

CHAPTER 4

Miniature Sorption Coolers - Part 1

THE FIRST closed-cycle sorption cooler was described by Torre and Chanin [88] and represented the first small self-contained system that achieved 350 mK for several hours with a recycling duty-cycle better than 90%. It was immediately clear that such systems, due to their compact, lightweight and self-contained properties, were well suited for applications where ease of operations and no external gas handling systems were useful. Astrophysical applications, for example, require cryogenic systems to be mounted on telescopes in remote sites where all the above characteristics are practically essential. Many more systems, with increased complexity and better performances, were built and used in many astrophysical experiments. We will describe some of the most used systems, from single-stage to triple-stage sorption refrigerators. We will then describe in detail how to properly design a single-stage and a double-stage sorption cooler. Finally, we discuss possible ways to achieve continuous operations with sorption coolers with completely sealed closed-cycle systems.

4.1 ADIABATIC LIQUEFACTION

Before discussing more in detail the functioning of sorption coolers, let's show how adiabatic transformations can be used to achieve cooling of a fluid. One of the formulations of the third law of thermodynamics is the Nernst theorem:

$$\lim_{T \to 0} \Delta S = 0 \tag{4.1}$$

which, in the case of low-temperature liquids, can be stated as "The entropy change associated with any condensed system subject to reversible isothermal process approaches zero as the temperature at which it is performed approaches 0 K". According to Mandl [54], we can state that a cooling

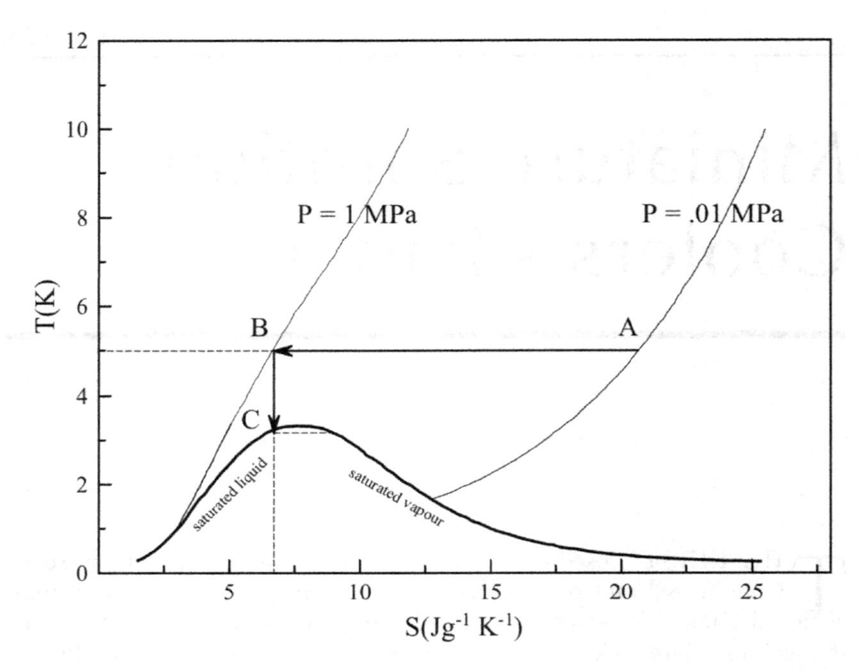

FIGURE 4.1: Adiabatic cooling with liquefaction of ^{3}He

process "squeezes" entropy out of a system. If the entropy of a system depends on the temperature and some other parameter β, we can write:

$$S = S(T, \beta) \tag{4.2}$$

Suppose we have an adiabatic process for which $S = S(T, \beta) =$ constant. During such a process we can find that the parameter β is changed in such a way that, to keep the entropy constant, the temperature needs to decrease. In fact, the entropy content of a system depends on two opposing phenomena: thermal motion tending to increase and the action of an ordering phenomenon like, for example, the ordering induced by an external magnetic field. If we are in a situation in the system where the transformation is isentropic, then if we release the ordering influence, for example by reducing the external magnetic field, the system tends to become more disordered. In order to keep the entropy constant, a counter-balancing decrease in temperature is necessary. Under the proper conditions, a single stage ^{3}He refrigerator can be shown to condense ^{3}He even with the condenser hotter than the critical temperature ($\sim$ 3.3 K).

In Fig. 4.1, the entropy/temperature (S/T) diagram for ^{3}He is shown. Two isobaric curves are shown corresponding to the two pressures $P_1 = 0.01$ MPa $=$ 100 mbar and $P_1 = 1$ MPa $=$ 10 bar. In a closed-cycle single-shot refrigerator, the heating of the charcoal pump produces an increase in the internal pressure of the gas. If the temperature at the condenser (as shown

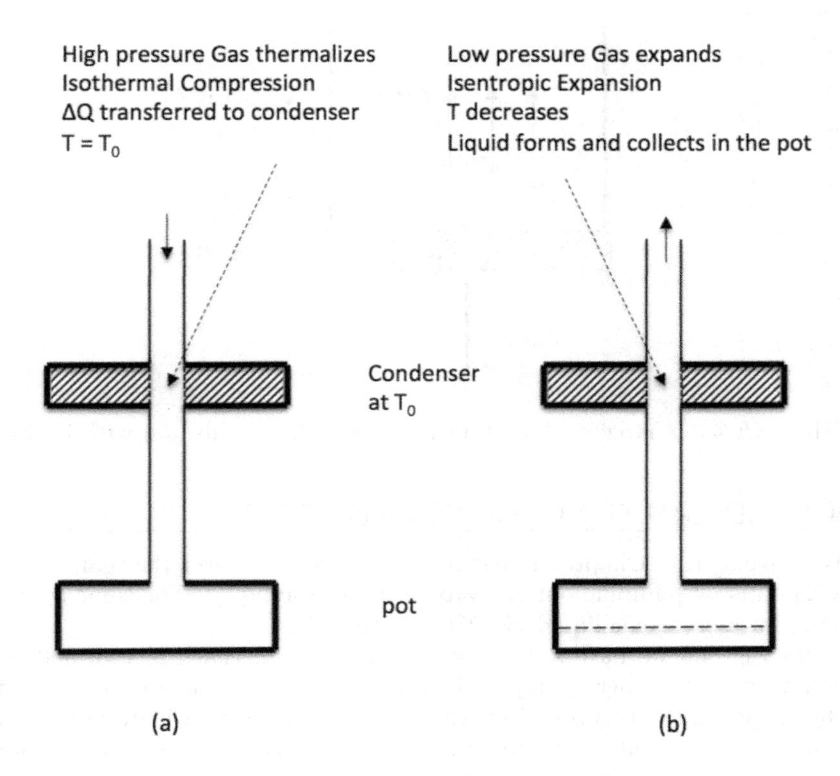

FIGURE 4.2: Adiabatic cooling with liquefaction of gas. (a) High-pressure gas is thermalized at the condenser. Heat ΔQ is transferred to the condenser. (b) Gas is expanded ($\Delta S = 0$) and temperature is decreased ($T_0 \rightarrow T_f < T_0$). If T_f is low enough, liquid can form and is collected in the pot even if T_0 is above the critical temperature.

in Fig. 4.2) does not change substantially during the pressure increase, we can assume that the compression is isothermal at the temperature of the condenser (process $A \rightarrow B$). Once the desired pressure has been reached, we can now turn off the heater to the charcoal pump. Activating the heat switch will produce a rapid cooling of the charcoal pump which, in turn, produces an isentropic adiabatic expansion (process $B \rightarrow C$). The expansion will bring the system onto the liquefaction curve and a fraction of the ^{3}gas will be liquefied even though the system has nowhere any temperature below the ^{3}He critical temperature (3.3 K). It has been shown [30] that with this type of cooling, it is possible to liquefy 46% of the initial amount of ^{3}He.

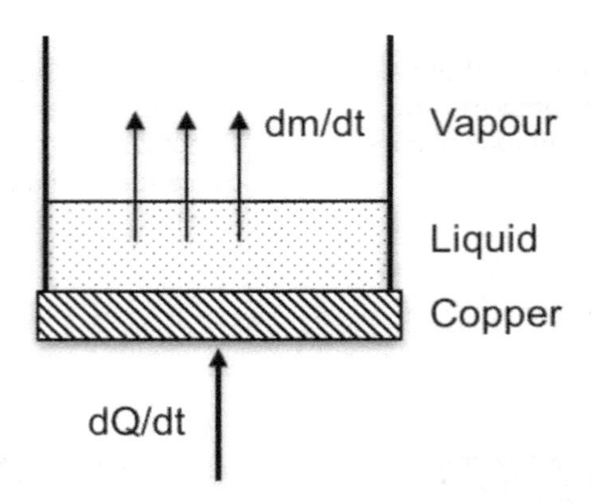

FIGURE 4.3: Cryogenic liquid inside a vessel in equilibrium with its vapour

4.2 GENERAL PRINCIPLES OF OPERATION

We have seen in Chapter 1 that a practical way to lower the temperature of a liquid is by pumping out the vapour above the liquid. The same concept is valid for a cryogenic liquid like ^{4}He or ^{3}He.

In Fig. 4.3 we have a cryogenic liquid inside a vessel in thermodynamic equilibrium. The thermal input dQ/dt to the liquid, coming for example from the supporting structures, electrical wires and radiation, will evaporate a fraction of the liquid at a rate dm/dt. The cooling action due to the evaporation will counteract the spurious heating action and will stabilize the system to a new, usually higher, temperature. The cooling power of a refrigerator is the function $T = T(\dot{Q})$ and is often quoted as the power to reach a typical temperature. For example, the cooling power of a G-M mechanical cooler is quoted to have 1.5 W at 4 K on its second stage, which means that if we apply 1.5 W on its second stage, the system will stabilize at an equilibrium temperature of 4 K.

We now express qualitatively the cooling power of our system depicted in Fig. 4.3. The equilibrium condition between the thermal input $\dot{Q}$ to the system and the cooling action due to the evaporation is

$$\dot{Q} = \dot{n}(H_l - H_{vap}) \tag{4.3}$$

where $\dot{n}$ is the number of molecules evaporating per second and $H_{l,vap}$ are the enthalpies of, respectively, the liquid and the vapour phase (per molecule). From the definition of latent heat L, we have:

$$\dot{Q} = \dot{n}L \tag{4.4}$$

the number of molecules evaporated per second is proportional to the vapour pressure at temperature T:

$$\dot{n} \propto P_{vap}(T) \tag{4.5}$$

It follows that

$$\dot{Q} \propto LP_{vap} \propto e^{-\frac{1}{T}} \tag{4.6}$$

The above equation shows that the cooling power decreases with T (rapidly). In order to have a powerful refrigerator, i.e., a refrigerator that does not increase too much its equilibrium temperature if we increase the thermal input, we need to keep the vapour pressure above the liquid as low as possible. This means that we need a powerful pumping action above the liquid bath. If we want to decrease the temperature of a pot containing, for example, liquid ^{4}He, we can pump over the liquid with a mechanical pump. In practical terms, it is quite difficult to reach temperatures in the pot below $\sim$ 1.3 K. If instead we use ^{3}He, we can reach lower temperatures of the order of $\sim$ 0.3 K. In Chapter 3 we have seen that it is possible to manufacture powerful pumps for He gases at low temperatures by using physisorption. It is therefore straightforward to imagine how we may realize a self-contained, closed-cycle refrigerator by coupling a charcoal pump to a pot containing liquid ^{4}He or ^{3}He. Fig. 4.10 below shows the basic layout of a ^{3}He sorption cooler mounted to a 4 K thermal reservoir. The first system built along these ideas was a single-stage ^{3}He system by Torre and Chanin [88] and it is schematically shown in Fig. 4.10. The critical parts of this system are essentially four: the cryopump (filled with sorbent materials), the condenser (in thermal contact with a source that allows the condensation), the evaporator (the pot that collects the cryogen) and the heat switch to cool/isolate the cryopump from the reservoir. Three more elements are important: the two tubes connecting, respectively, the cryopump to the condenser and the tube connecting the pot to the condenser.

4.2.1 The cryopump

We have seen in Chapter 3 the physics of sorption cryopumps, based mainly on physisorption, i.e., the ability to bound helium molecules to the surface of, for example, activated charcoal. In the rest of the book, we will mainly consider activated charcoal as the main absorbent compound inside the cryopumps.

Activated charcoal has two main behaviours depending on the physical temperature at which it is thermalized: (1) pumping phase for temperature $T < T_d$ and desorption phase for temperature $T > T_d$. T_d is a typical temperature of transition between these two regimes and, depending on the geometry and the construction of the cryopump, is of the order of 10 K $< T_d <$ 15 K. From Fig. 3.3 we see that if we manage to cool the cryopump to a base temperature of 4.2 K, then we have practically total absorption of the helium gas (no pressure dependence on the amount of gas absorbed per gram of charcoal). If, on the other hand, we heat the charcoal above 28 K, there are four orders of magnitude less helium absorbed in the charcoal: the helium is practically totally desorbed. Therefore, if we are able to heat/cool the charcoal inside the cryopump, we can switch between a pumping and a desorption phase. This is easily achieved by connecting a heater (usually a resistor in thermal contact with the charcoal) to the pump and connecting the pump to the cold base-plate through a heat switch. When we want to heat up the charcoal pump and desorb the helium, we turn the heat switch off (low thermal conductivity to the base plate) and turn on the heater until we reach a charcoal temperature $T >\sim 30$ K. When we want to pump and absorb the helium gas into the cryopump, we turn off the heater and turn the heat switch on (high thermal conductivity).

A good design of a cryopump will make sure that the charcoal inside the pump is in good thermal contact with the *hot side* of the heat switch. In order to achieve this result, it is possible to design a cryopump as shown in Fig. 4.4. In Fig. 4.5, it is possible to notice only the copper screws from the bottom of the cryopump. These screws allow to distribute vertically the heat relying on the high thermal conductivity of the copper and not on the contact between the different grains of charcoal. Instead, Fig. 4.5 shows a copper disk (spacer) that allows a radial distribution of the heat. This disk is in contact with the screws. Therefore, the strategy is to use multiple screws to distribute vertically the heat and multiple disks (every 2 cm) to distribute radially the heat. With this solution, the charcoal is close to be isothermal at the temperature of the bottom of the cryopump. In Figs. 4.4 and 4.5, it is possible to notice also a stainless steel tube coming from the bottom of the cryopump. This tube is holed and the dimension of each hole is inferior to the average diameter of a charcoal grain. In this way the gas can flow up to the top of the cryopump and the grains cannot fall downwards.

4.2.2 Helium pot

The helium pot is the chamber where the liquid helium is collected, usually by the action of gravity. The bottom of it needs to be manufactured out of a high thermal conductivity material so that the internal side (in contact with the liquid) and the external side have the smallest discontinuity due to the Kapitza resistance. A suitable material is Oxygen Free High Conductivity (OFHC) copper for the bottom. The sides and the top should be made from stainless steel to minimize the conduction from the tubes that connect with the warmer condenser to the liquid which is not in contact with the top. Below 1 K, the Kapitza resistance between liquid and metal is important and can introduce a ΔT up to 10% of the liquid temperature. It is important to reduce this ΔT because we need to remember that for any application only the external side of the pot will be used. A common solution to do that is using a series of concentric fins coming out from the base of the pot as shown in Fig. 4.6. These will increase the surface contact area between the base of the helium pot and the liquid, reducing the Kapitza resistance. Other solutions

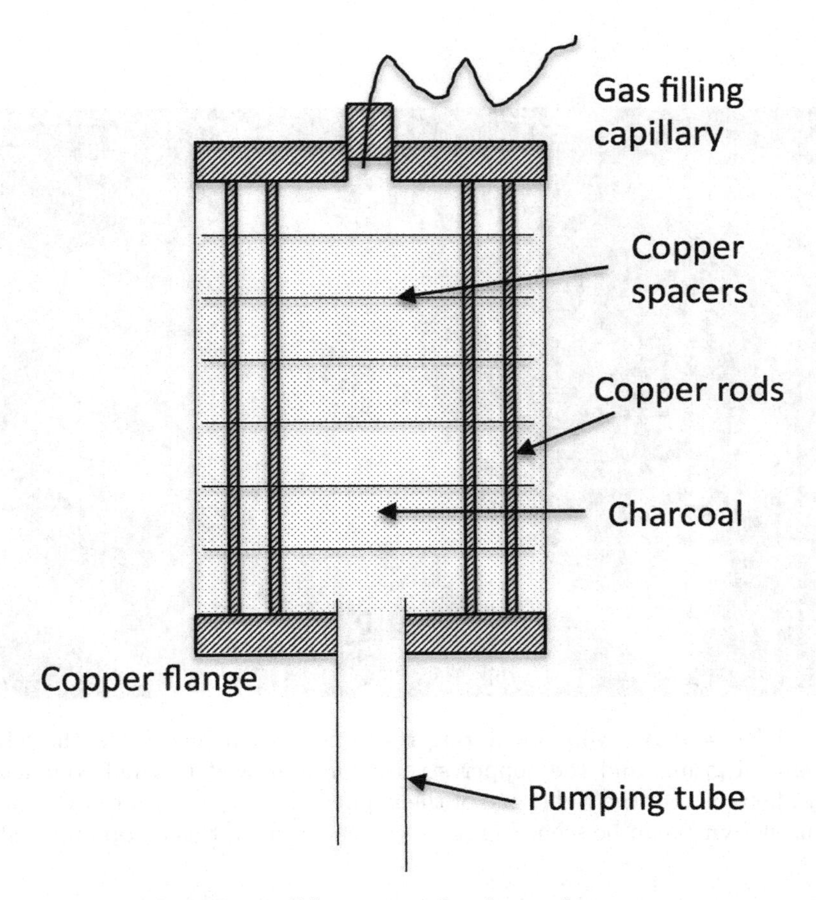

FIGURE 4.4: Schematics of a charcoal cryopump

FIGURE 4.5: A cryopump during the construction phase. On the left, the charcoal grains and the copper struts (for improved thermal conductivity) can be seen. On the right, one of the copper disks (also for improved thermal conductivity) can be seen. In Fig. 4.4 a schematic of this cryopump is shown.

FIGURE 4.6: Bottom of a helium pot with fins to reduce the Kaptiza resistance

are possible: sintered copper or silver can be used to increase the effective surface area of contact between liquid and metal.

4.2.3 Pumping tubes

Pumping tubes normally connect stages at different temperatures. The main requirements therefore are maximum strength and minimal thermal conductivity. Thin-walled stainless steel tubes are the common choice down to a few hundred mK. Below 100 mK, Cupro Nickel might be used. The performance of the refrigerator depends highly on the proper sizing of the pumping tubes and we will describe later in this chapter how to calculate the tubes.

4.2.4 The condenser

The condenser is the place where the helium gas is subject to condensation and, once it becomes liquid, it is pulled down by gravity into the helium pot. The condenser needs to be in very good thermal contact with the stage that is providing the condensation temperature. In addition, it has to provide minimal drop in pressure not to impact the pumping phase over the liquid helium bath. The preferred material is OFHC copper. At very low temperatures, Kapitza resistance becomes important and improved contact area solutions need to be implemented. When the ^{4}He is the gas to be condensed, it is imperative to have the condenser at the lowest possible temperature and with the best thermal contact. The efficiency of condensation depends on various physical and geometrical factors but it is also a strong function of temperature.

In the case of a closed-cycle sorption cooler, if we assume that the ^{4}He is an ideal gas, then the pressure at any temperature T, assuming that the system is isothermal, is given by:

$$p(T) = \frac{p(300)T}{300} \tag{4.7}$$

where $P(300)$ is the pressure inside the refrigerator at $T = 300$ K. We want to be sure that when our charcoal pump is at $T \sim 30$ K the internal pressure is exceeding the saturated vapour pressure at the condensation point. Since typical room temperature pressures of sealed systems are in the range $80 \div 100$ bar, we see that these systems easily exceed the saturated vapour pressure and condensation is ensured.

Obviously at low temperatures the gas is not ideal and there is no isothermal condition within the system. The condensation itself will reduce the internal pressure so Eq. 4.7 is not a good representation of the system. We must apply a more sophisticated analysis of the system accounting for geometry and the temperature distribution; this is detailed in Section 4.2.7.

For now, however, let us consider only the effect of the condenser temperature on the condensation efficiency of the fridge. We take here as an illustrative example the 1 K sorption cooler developed by the authors for the QUBIC experiment [59, 5, 56] (the system itself is shown in Fig. 4.7 and the geometrical parameters are given in Table 4.1).

Upper tube length	153 mm
Upper tube radius	16 mm
Upper tube wall thickness	0.91 mm
Lower tube length	56 mm
Lower tube radius	10 mm
Lower tube wall thickness	0.25 mm
Volume of charcoal space	$3.10\ 10^5$ mm^3
charcoal filling factor	0.64
Effective volume of charcoal space	$1.98\ 10^5$ mm^3
Upper tube volume	$1.22\ 10^5$ mm^3
Lower tube volume	$1.59\ 10^4$ mm^3
Volume of pot	$1.81\ 10^5$ mm^3
Total volume of SC	$4.3\ 10^5$ mm^3

TABLE 4.1: ^{4}He QUBIC fridge geometrical specifications

Using the analysis in Section 4.2.7 and assuming a typical gas charge (again detailed in Section 4.2.7), the resulting efficiency for both ^{4}He and ^{3}He is shown in Fig. 4.8.

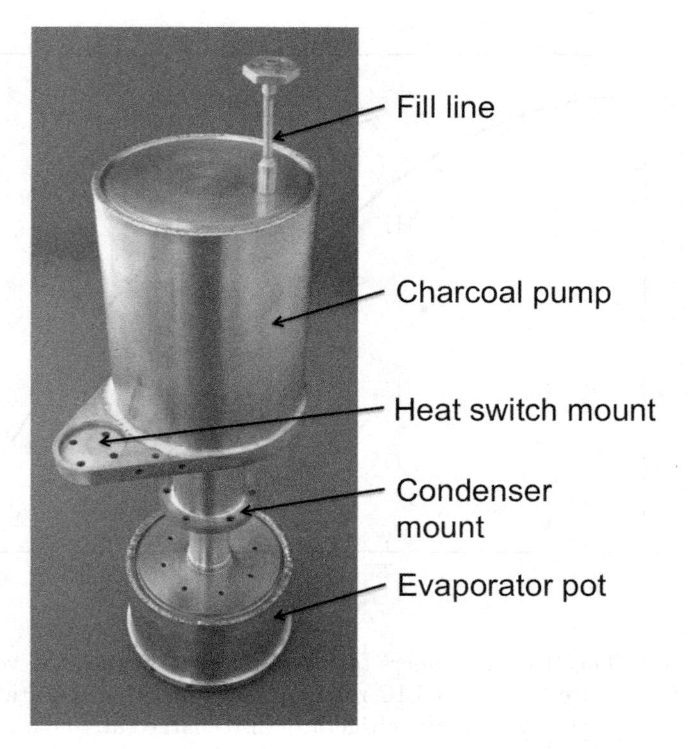

FIGURE 4.7: The ^{4}He fridge developed for QUBIC [56]

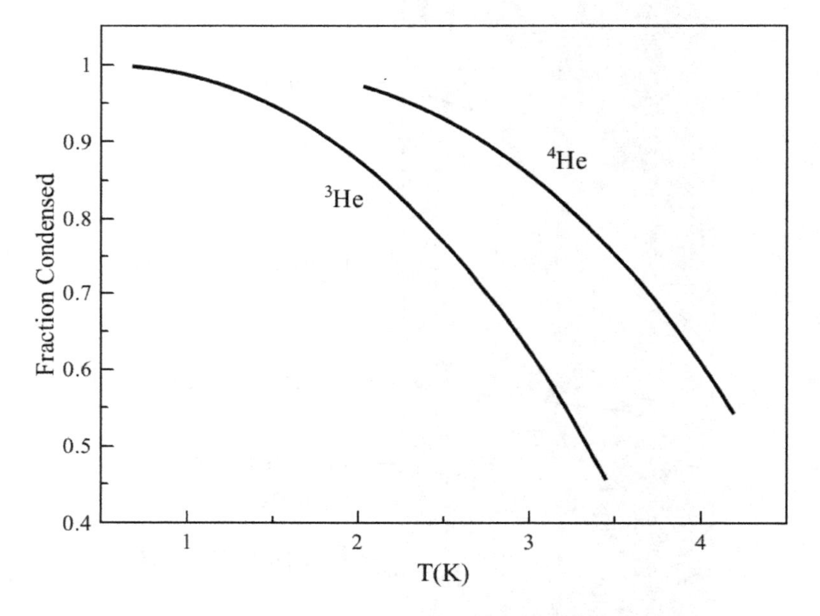

FIGURE 4.8: The ^{4}He curve refers to the condensation efficiency versus condenser temperature for the QUBIC refrigerator charged at 70 bar at $T = 293$ K. The ^{3}He curve refers to a ^{3}He refrigerator also charged at 70 bar at $T = 293$ K. [59]

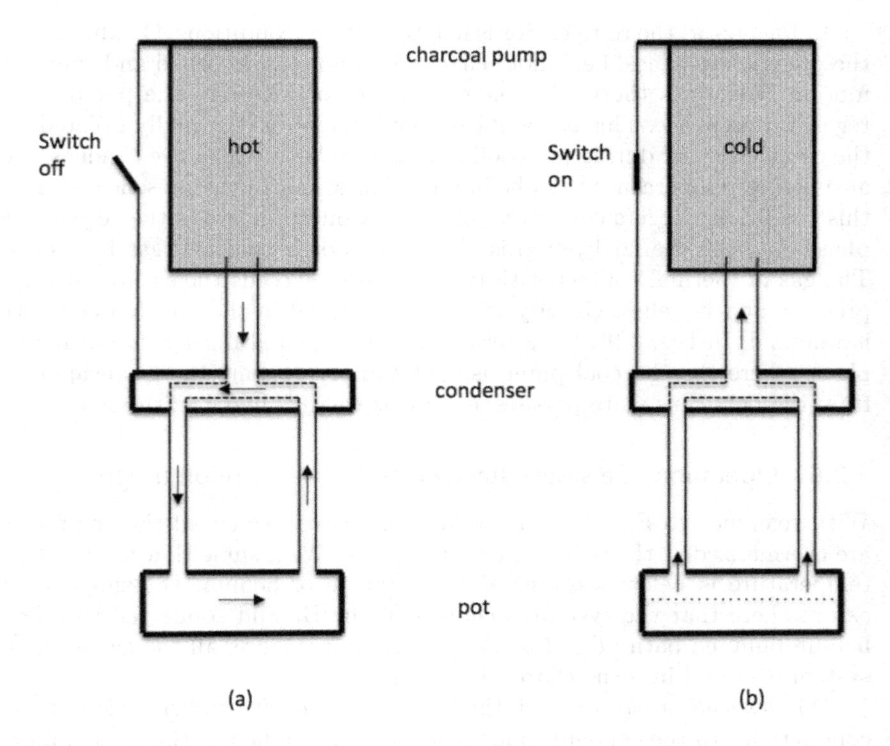

FIGURE 4.9: Single-stage refrigerator with two pumping tubes in parallel to allow convective pre-cooling. (a) The refrigerator is either in the regeneration or pre-cooling phase. (b) The fridge is in the cold phase. Note the status of the heat switches and the gas circulation indicated by the arrows.

4.2.5 Heat switch

The main characteristics of a good heat switch are fast on/off switch time (at least compared with the typical time scales of the sorption fridges) low "off" thermal conductivity and high "on" thermal conductivity. Several techniques can be employed, from superconducting wires to mechanical switches. The preferred choice in sorption coolers is to use either gas-gap or convection heat switches. In these kinds of switches the off conduction is dominated by the stainless tube, while the on conduction is dominated by the thermal conductivity of the helium gas. On/off ratios in excess of a few hundred can be reached with relatively easy designs.

Heat switches are used for two reasons: pre-cooling of the cold stage and connection of charcoal pumps to the main reservoir. In some designs [60, 17], the pre-cooling can be achieved by incorporating a convection circuit in the pumping lines of the refrigerator. An example of such configuration is shown in Fig. 4.9.

In Fig. 4.9(a) the refrigerator is in either of two conditions. Condition one is the pre-cooling phase, i.e., "hot" charcoal pump, gas desorbed and convective motion of the gas thermally connecting the condenser to the pot or in the regeneration phase. This is possible because the pot is thermally isolated from the condenser and during pre-cooling it is both hotter than the condenser and at a higher temperature than helium condensation. In the presence of gravity this condition triggers convective motions. Condition two is the regeneration phase, i.e., the charcoal pump is "hot," the pot is cold and gas is desorbed. The gas in thermal contact with the condenser exceeds the saturated vapour pressure and liquefies. Gravity will pull the liquid in the pot and convection is minimal. In Fig. 4.9(b) the refrigerator is depicted during its normal cold phase where the charcoal pump is cold and cryosorbing the gas evaporating from the pot. This state persists as long as there is liquid in the pot.

4.2.6 Operation of a single-stage closed-cycle sorption fridge

With reference to Fig. 4.10, a cooling cycle starts when all the components are thermalized at the base-plate temperature. We assume that the condenser temperature is below the critical temperature of helium. For simplicity we assume here that the system is charged with ^{3}He and connected to a liquid helium pumped bath ($T \sim 1.5$ K). At this temperature, all the gas inside the system is bound into the charcoal cryopump.

(1) *Desorption phase*: With the heat switch in *off* condition (low thermal conductivity to the condensation stage), heat is applied to the cryopump and the charcoal inside is heated to about 40 K where the gas is practically all released. In this condition the internal pressure is high enough to exceed the the saturated vapour pressure ($\sim$ 70 mbar for a temperature of 1.5 K).

(2) *Condensation phase*: During this phase, the temperature of the cryopump is regulated to be constant while the gas is completely desorbed. Liquid is constantly produced at the condenser and is collected in the pot. The temperature of the pot will stabilize to a temperature slightly higher than the condenser temperature and it can be noticed that the amount of power needed to keep constant the temperature of the cryopump decreases as the condensation progresses. The condensation phase typically takes 10 to 20 minutes, depending on the size of the cryopump or the amount of gas to be condensed.

(3) *Pumping phase*: Once the condensation is completed, power to the cryopump heater can be removed and the heat switch can be turned on. This action will rapidly begin the pumping action over the liquid bath in the pot. The temperature is observed to decrease until it reaches the minimum temperature when the cooling power of the liquid helium bath is equal to the external thermal input power.

(4) *Regeneration phase*: Once all the liquid is pumped back into the cryopump, the cooling power in the pot goes rapidly to zero and the various spurious thermal inputs will heat up the pot back to the base-plate stage temperature. At this point, if necessary, a new cycle can be started.

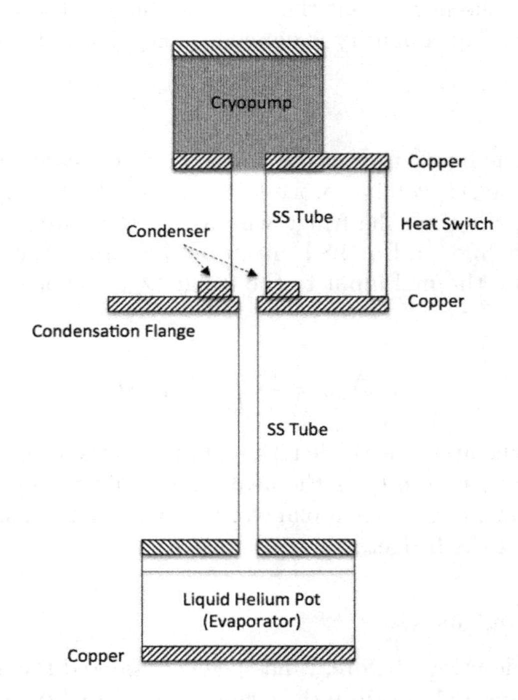

FIGURE 4.10: Basic scheme of a single-stage, closed-cycle sorption cooler

The process just described can be divided in two main parts: the cycling phase (the phase composed by the liquefaction of the gas and the evaporation until the final temperature is reached) and the operative phase (the phase where the liquid is at constant temperature in the evaporator). The last phase is the most important one since it is the phase when the liquid is able to cool down everything that is thermally connected to the pot. At sub-K temperatures the specific heat of liquid helium is much larger than the specific heat of metals. Therefore, even several kilograms of metals attached to a few cm^3 of liquid helium will follow the temperature of the liquid.

The most important characteristic of a sorption cooler is the *cooling power*, which describes the power that the fridge is able to absorb without changing its temperature. This quantity is defined by Eq. 4.4 and can be written as:

$$\dot{Q} = \frac{Ln}{t} \tag{4.8}$$

where L is the latent heat of the liquid, n is the quantity of the liquid that remains after the self-cooling process is completed and t is the hold time of the fridge (i.e., the time the fridge will last at the lowest temperature). The cooling power defined in Eq. 4.8 is given by the sum of the external load $\dot{Q}_{ext}$ and the parasitic thermal input to the fridge $\dot{Q}_{par}$. This is a convective term, so it is equal to

$$\dot{Q}_{par} = \frac{A}{l}\int_{T_e}^{T_c} k(T)dT \tag{4.9}$$

where A is the area, l is the length and $k(T)$ is the thermal conductivity of the tube. The other term $\dot{Q}_l$ is the heat load coming from the object that we want to cool down (e.g., a focal plane or a super-magnet), so it is the effective cooling power of the fridge.

4.2.7 Charging gas

In the process described before, it has been considered the ideal case when all the gas condenses and falls into the evaporator. In reality, it happens that not all the gas becomes liquid, so the *condensation efficiency* of the system is always lower than 100%. We have already shown the condensation efficiency of a single-stage ^4He refrigerator (QUBIC) based on a specific geometry. Let's now estimate in general the condensation efficiency. Let's write the total number of moles of gas in the system:

$$n = n_l + n_e + n_t + n_p \tag{4.10}$$

where the subscripts l, e, t and p mean, respectively, the fraction of gas in liquid form in the evaporator, in gas form in the evaporator, in gas form in the tubes and in gas form in the cryopump. In order to compute these quantities, we assume (1) that the pressure in the fridge is constant and equal to the pressure of the gas at the condensation temperature, (2) that the temperature of the evaporator and the pump can be described by a single value for each component and (3) that the profile temperature through the tubes is linear. Under these assumptions we can compute the following quantities:

- $n_e = P \cdot (V_e - n_l\tilde{V})/RT_e$, where $\tilde{V}$ is the molar volume that allows to consider that not all the evaporator is filled with gas, but there is also a liquid,

- $n_p = PV_p/RT_p$,

- $n_t = \pi P/R \cdot (r_{ec}^2 \int_0^l dz/T(z) + r_{cp}^2 \int_l^L dz/T(z))$, where the first term is referred to the tube from the condenser to the evaporator and the second to the tube from the condenser to the pump.

It may then be seen that by equating the pressures in the different sections of the system, we get the number of moles of liquid as

$$n_l = \frac{n - n_t - \frac{p}{R}\left[\frac{V_p}{T_p} + \frac{V_z}{T_z}\right]}{1 - \frac{p\tilde{V}}{RT_p}} \tag{4.11}$$

Further to the initial condensation efficiency loss, another efficiency loss occurs when initially pumping as the helium self cools from the condensation temperature to the base equilbrium temperature. To compute the quantity of liquid evaporated due to this process, we need to estimate the heat needed to cool n moles of gas of an amount dT:

$$nC(T)dT = L(T)dn \tag{4.12}$$

where $C(T)$ is heat capacity of the liquid and $L(T)$ is the latent heat, both functions of temperature. With all the formulas given above, condensation efficiency can be now estimated with some accuracy.

4.2.8 Pumping speed

An important factor to reduce the vapour pressure over the liquid is the ability to remove efficiently the evaporated gas atoms, i.e., we need to have a pump able to remove quickly a large amount of gas and leave a minimal residual pressure. It is well known that even if we use a very powerful pump, extra care must be taken in order to reduce the pressure drops through the pumping lines. In our case, the tubes that allow the flow of the gas from the pot to the cryopump must not limit the performances of the pump. More specifically, they need to have high conductance. In vacuum technology, the conductance G is defined as:

$$G = \frac{q_{PV}}{\Delta P} \tag{4.13}$$

where ΔP is the pressure differential at the edges of the tube and q_{PV} is the so-called *volumetric throughput*, defined as $q_{PV} = PdV/dt$, where dV/dt is the volume of gas per unit of time that crosses a perpendicular plane to the flow and P is the pressure where the crossing is measured.

The volumetric throughput (also called PV-throughput) is obtained by dividing Eq. 1.59 by time:

$$q_{PV} = \frac{PV}{t} = \frac{mRT}{Mt} \tag{4.14}$$

which shows that, at constant temperature, the mass flow is constant.

Eq. 4.13 is formally equivalent to Ohm's law $1/R = I/V$, where $1/R$ is the electrical conductance, and I, V are, respectively, the current and the voltage. Pushing the analogy a bit further, we can then identify the electrical conductance $1/R$ with the conductance G, the voltage V with the pressure difference ΔP and the current I with the flow q_{PV}. We immediately have formulas to calculate the conductance of pumping lines in parallel and in series:

$$G_{parallel} = G_1 + G_2 + \ldots + G_n \tag{4.15}$$

$$\frac{1}{G_{series}} = \frac{1}{G_1} + \frac{1}{G_2} + \ldots + \frac{1}{G_n} \tag{4.16}$$

In cryogenic application it is more convenient to use the mass flow G_m defined as follows:

$$G = k_B T \dot{N} = G_m \frac{k_B T}{m_m} \tag{4.17}$$

where k_B is the Boltzmann constant, $\dot{N}$ is the number of atoms per unit of time crossing the plane and m_m is the molecular mass.

For cylindrical section tubes of radius r with length $l \gg r$, the conductance is

$$G = \frac{4r^3}{3l} \sqrt{\frac{2\pi k_B T}{m_m}} \tag{4.18}$$

When the condition $l \gg r$ is not satisfied, we write the conductance as

$$G_r = \frac{3r}{8l} K_f G \tag{4.19}$$

where K_f is the *Clausing Factor* that takes into the account the non-ideal condition of the tube and defined as

$$K_f(x) = \frac{0.98441 + 0.00466x}{1 + 0.46034x} \tag{4.20}$$

where $x = l/r$.

Combining Eqs. 4.13, 4.17 and 4.19, it is possible to express the mass flow as

$$G_m = \frac{Cr^2 K_f(x)}{T} \Delta P \tag{4.21}$$

where $C = \sqrt{\pi m_m/(2k_B T)}$.

In this case, the temperature is constant through all the tube. However, for a sorption cooler this formula is not enough, because the tube connects two parts with a different temperature. In this case, Eq. 4.21 becomes

$$G_m = Ar^2 K_f(x) \left(\frac{P_1}{\sqrt{T_1}} - \frac{P_2}{\sqrt{T_2}} \right) \tag{4.22}$$

If we consider the mass flow constant, it is possible to include the term G_m into the constant C, creating a new constant $B = G_m/C$, so Eq. 4.22 becomes

$$\frac{B}{K_f(x)r^2} = \left(\frac{P_1}{\sqrt{T_1}} - \frac{P_2}{\sqrt{T_2}} \right) \tag{4.23}$$

In general, for a pumping system containing tubes with different dimensions, we have

$$B \sum_{i=0}^{n} \frac{1}{K_f(x_i) r_i^2} = \left(\frac{P_0}{\sqrt{T_0}} - \frac{P_n}{\sqrt{T_n}} \right) \tag{4.24}$$

For a sorption cooler the subscript 0 refers to the coldest point (pot) and the subscript n to the hottest point (cryopump). In this case, since the pressure in the cryopump is significantly lower (the atoms are adsorbed by the sorbent) and the temperature is comparable, it is possible to neglect the last term, so

$$B \sum_{i=0}^{n} \frac{1}{K_f(x_i) r^2} \leq \frac{P_0}{\sqrt{T_0}} \tag{4.25}$$

This last equation describes the criteria that the pumping line of a sorption cooler should respect.

4.3 1 K SYSTEMS

A 1 K sorption cooler is basically the same system described in the previous section, in which the gas is ^{4}He.

This system can reach temperatures below 1 K. However, as shown in Fig. 1.2, the value of the vapour pressure drops quickly decreasing the temperature and all the cryopumps struggle to create a vacuum with a pressure less than 10^{-3} mbar, which corresponds to a minimum temperature just below 700 mK. This means that these systems at low temperatures are pump limited. Typical cooling power of these systems spans from hundreds of µW to some mW.

These systems are usually coupled with a mechanical cooler that provides a temperature low enough to allow the condensation. Usually, this temperature is in a range between 2.8 and 4.2 K. The condensation temperature is fundamental during the designing phase of any sorption cooler. Indeed, the gas, below the condensation temperature, condenses only if pressure is greater than the vapour pressure. As an example, this means that the pressure at 4.2 K should be greater than 0.99 bar, which is the vapour pressure at this temperature. Therefore, when a sorption cooler is charged there is the necessity to take into consideration that the quantity of gas should be enough to compensate (partially) the decreasing pressure due to the decreasing temperature thanks to the mechanical cooler.

Another key element that needs to be considered is the presence of superfluid ^{4}He below the lambda transition; note that this only occurs with ^{4}He (and not ^{3}He) due to the isotope having integer spin and therefore obeying Bose-Einstein statistics as described in Section 1.2.2.

4.3.1 Superfluid film breaker

Some behaviour of superfluid helium can be described by the two-fluid model of Landau and Tisza, which treats Helium-II as being composed of a normal and superfluid component with their own independent sets of properties [46]. The equation describing the acceleration of the superfluid component [95] is

$$\frac{\partial v_s}{\partial t} = -\frac{1}{\rho}\nabla P + s\nabla T \tag{4.26}$$

from which it may be seen that (a) the superfluid flows under the influence of pressure gradients and (b) that it also responds to gradients in temperature. It may be understood from the so-called "fountain" term $s\nabla T$ that the superfluid will accelerate towards warmer regions giving rise to the thermomechanical or "fountain pump" effect.

It may be seen therefore how, for ^{4}He coolers, superfluidity introduces additional complications. As a film will tend to climb the inside of the pot, liquid may be drawn out of the pot and into the pumping tube where it evaporates without producing useful cooling (reducing the hold time) and further the presence of the film may increase the thermal load on the evaporator (raising the equilibrium temperature).

In principle, several solutions may be suggested to mitigate this effect. A heater may be employed to raise the temperature of a point on the pumping tube sufficiently high so as to break the superfluid state at the top of the pot; this, however, necessarily introduces a significant additional heat load to the pot and is undesirable for this reason. A second option may be to employ some surface in the construction of the top of the evaporator pot over which, for whatever reason, superfluid is unable to flow or at least is significantly impeded; unfortunately, however, the literature shows little difference (~25%) in film flow rate over available materials [77, 19] or any significant variation with surface finish [78]. A final option is to introduce some geometrical feature to impede the flow of the film and it is with this approach that the following solution has been found.

This film breaker consists essentially of a short capillary-like constriction placed at the top of the evaporator, which is designed to restrict the film flow whilst not significantly impeding the gas flow [58]. A knife edge film breaker of similar construction was also reported by Lau et al. [47].

In order to pass through the restriction (in the direction of the positive temperature gradient), the film has to flow over a sharply machined edge. As given by Shirron and DiPirro [75], the limiting film thickness d and radius of curvature r_s over which the film has to flow are related by

$$d^3 = \frac{\Gamma}{k_B T \ln\left(\frac{P_{vap}}{P}\right) - \frac{m\sigma}{\rho r_s}} \tag{4.27}$$

where Γ is the van der Waals constant, m is the mass of the helium atom, σ is the surface tension and ρ is the superfluid density. As such, introducing an extremely sharp edge into the film flow path (having a small, negative radius of curvature) results in a large surface tension energy, reducing the thickness appreciably. It is clear from continuity that this limits mass transport through the orifice, as the film is also bounded by a critical velocity [95].

In order to simultaneously not restrict the gas flow, the orifice is designed to choke the flow passing, hence reaching the sonic condition and therefore maximising mass transport. Considering the analagous case of a convergent-divergent rocket nozzle from aerodynamics, it may be seen that where the pumping line downstream of the constriction is sized suitably that the back pressure is below the critical pressure, the sonic condition is reached at the orifice. The mass transport is then

$$\frac{dm}{dt} = \pi r^2 \left(\frac{p_0 M}{TR}\right)\left(\frac{\gamma TR}{M}\right)^{1/2}\left(\frac{1}{2\pi\gamma}\right)^{1/2} \tag{4.28}$$

in the molecular flow regime and

$$\frac{dm}{dt} = \gamma^{1/2}\left(\frac{2}{\gamma - 1}\right)^{\frac{(1/2)(\gamma+1)}{\gamma-1}} \pi r^2 p_0 \left(\frac{M k_B}{T}\right)^{1/2} \tag{4.29}$$

in the continuum regime. Both models assume that the gas is ideal, the specific heats at constant pressure and constant volume vary sufficiently little with temperature in this regime that they may be considered constant, the process is adiabatic and that gravitational effects can be neglected. As seen later from the agreement of experimental data with the model, these are held to be reasonable assumptions.

The model is somewhat complicated by the fact that, as may be calculated from the vapour pressure curve and taking the characteristic length as the restriction diameter, the system is in the transition regime at lower temperatures and then approaches the continuum regime at higher temperatures.

Fig. 4.11 below shows the models for both flow conditions and it may be seen that at lower temperatures (i.e., in the transition regime) the experimental data lie bounded by the models, and as the temperature increases the data tend to the continuum model.

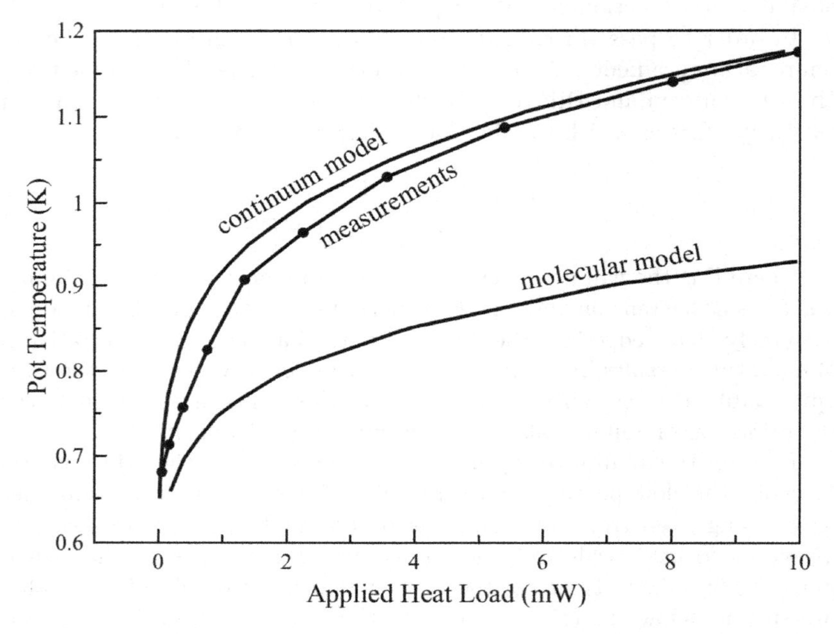

FIGURE 4.11: Restriction flow models with experimental data for 750 μm diameter restriction (line added to measurements to guide the eye)

The model is therefore strongly predictive at higher temperatures and may serve as a conservative guide at lower temperatures. Extensive experimental tests have shown this form of film breaker to allow negligibly small film flow rates (within the range of experimental error) and accordingly no measurable reduction in hold time.

An example code, written in Python, to model the load curve for a given film breaker geometry is given in the Appendix in Section B.1.

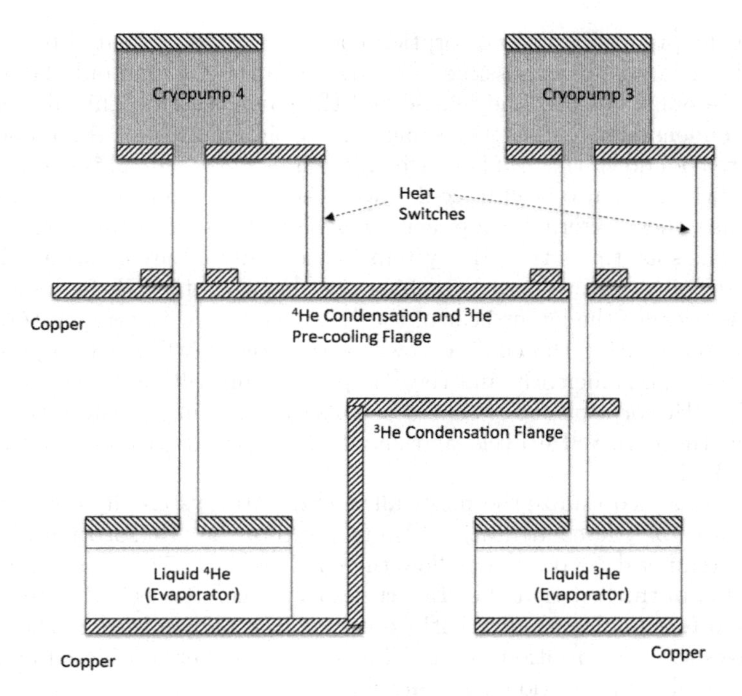

FIGURE 4.12: Basic scheme of a ^{3}He/^{4}He sorption cooler

4.4 300 MK SYSTEMS

The main difference between these systems and the previous ones is due to the fact that the 300 mK systems use ^{3}He instead of ^{4}He. The working principles of these systems are mainly the same as the previous ones.

These coolers are able to reach temperatures below 300 mK. This is mainly due to the fact that the vapour pressure of ^{3}He is higher than that of ^{4}He. Indeed, the sorbent continues to create the same vacuum but the difference in the vapour pressure curve between these two isotopes allows ^{3}He systems to reach lower temperatures than ^{4}He. Usually, the cooling power of these system is on the order if tens of μW.

Compared to ^{4}He, ^{3}He has a lower condensation temperature (at the same pressure), meaning that many of the mechanical coolers typically used in many research laboratories are unable to condense liquid as required in a ^{3}He system. However, it is instead possible to use a ^{4}He cooler in tandem with a ^{3}He cooler, where the first acts as a precooler to condense liquid in the second, as well as intercepting the heat load through the pumping tube from the 4 K plate. This solution is widely used nowadays and such a system is presented in Fig. 4.12. Singularly both the ^{3}He and the ^{4}He sorption coolers work as explained above, but together there are some minimal differences.

In order to use such a system, there is the necessity to optimize the time when the pumps of the two sorption coolers start to be heated to release the gas. For example, it is possible to condense all the ^{4}He and only later heat up the ^{3}He pump to start the release and the condensation. This allows a quick ^{4}He condensation, but at the same time a big quantity of the liquid will be used to cool down the ^{3}He later when it will be desorbed. This can be useful if the ^{4}He stage is also used to cool down other parts (such as wires). Indeed, in this case, these external components reach quickly the operative temperature. However, sometimes the ^{4}He system is used only to pre-cool, in this case it is more convenient to start desorbing the ^{3}He when the ^{4}He is desorbing too. This is because the latent heat of the ^{4}He decreases with temperature, so this means that during the cooling down of the ^{4}He, it will start to pre-cool the ^{3}He gas[1] more efficiently and the ^{3}He pre-cooling will be driven by the ^{4}He.

The ^{4}He sorption cooler should be able to maintain a constant temperature during the entire ^{3}He cycle and should be charged with a quantity of gas accordingly.

In order to optimize thermally all the ^{3}He/^{4}He system, it is possible to use the source of ^{4}He condensation also to pre-cool the ^{3}He sorption cooler, also if this temperature does not allow the condensation. The reason for this can be found in the fact that the ^{3}He arrives to the condenser with a temperature below 5 K, significantly low with respect to the temperature of the gas when it leaves the condensation pump. This may be further considered by referring to the analysis in Section 4.2.7 and Fig. 4.8.

4.5 220 MK SYSTEMS

The systems described in the previous sections are limited by the heat load from the condenser through the tube. This is usually tens of µW, so comparable with the external load on the system. If we consider that this load becomes worse trying to go down in temperature, it is easy to understand that the previous system is limited by this factor. In order to solve this problem, it is possible to add another ^{3}He sorption cooler to the ^{3}He/^{4}He system. In this way, the ^{3}He stage of the ^{3}He/^{4}He fridge is used to supply a 300 mK to allow the low-temperature condensation of ^{3}He in the fridge. This system is now composed of two ^{3}He sorption coolers and one ^{4}He; its schematic is presented in Fig. 4.14.

Usually, the intermediate stage, the one that provides 300 mK, is used only to condense the ^{3}He for the final stage. To fully optimize thermally this system, the condensation stage for the ^{4}He is used to precool both the ^{3}He sorption coolers. Moreover, the 1 K flange is used not only to condense the ^{3}He in the intermediate stage, but also to pre-cool the ^{3}He in the final stage. Consequently, the ^{4}He stage should be able to provide enough cooling power to the ^{3}He in the intermediate sorption cooler and pre-cool the ^{3}He in the

[1]It may not condense if the condensation temperature of the ^{4}He is higher than 3.2 K.

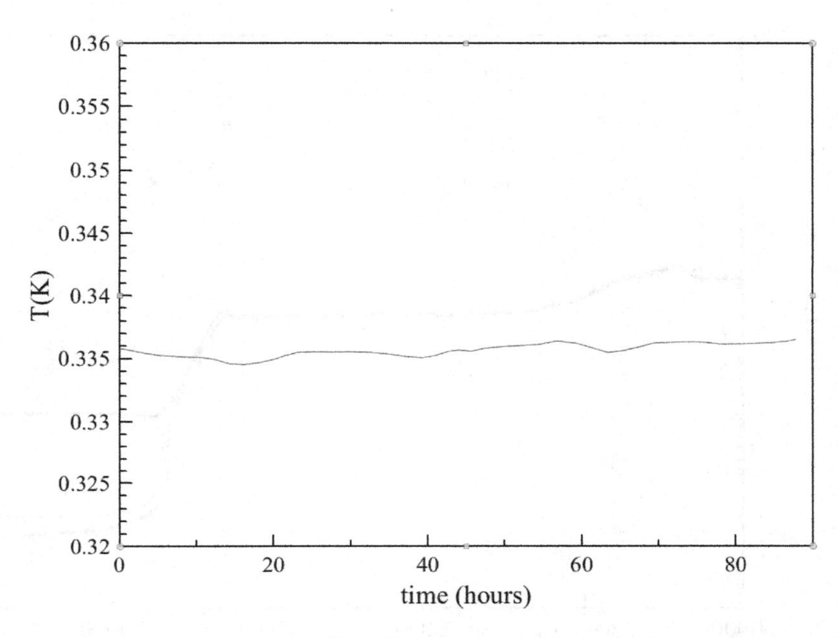

FIGURE 4.13: Hold time of a ^{3}He/^{4}He sorption cooler with a load of 20 μW. Temperature fluctuations of the cold stage are $\pm$ 1 mK.

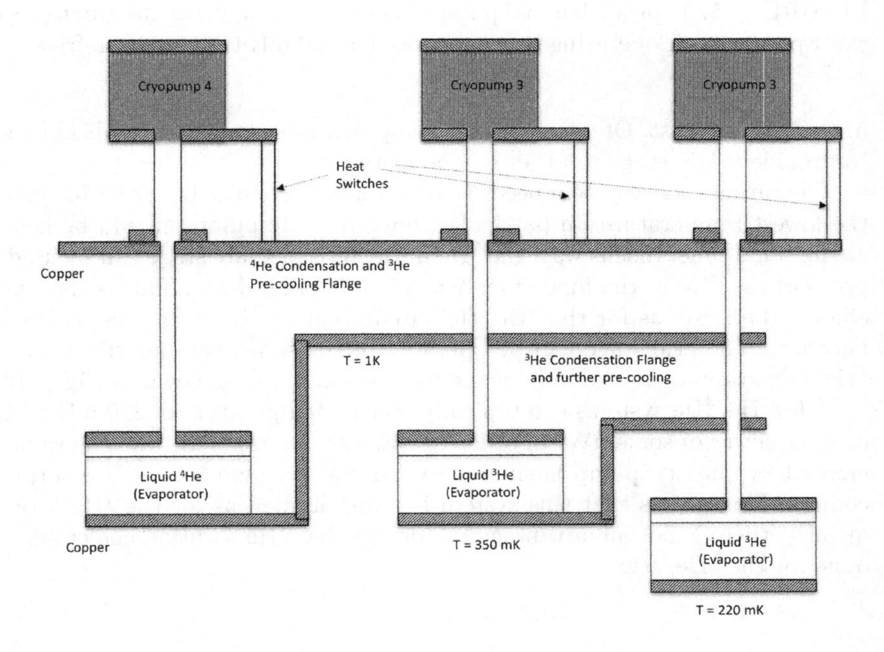

FIGURE 4.14: Basic scheme of a ^{3}He/^{3}He/^{4}He sorption cooler

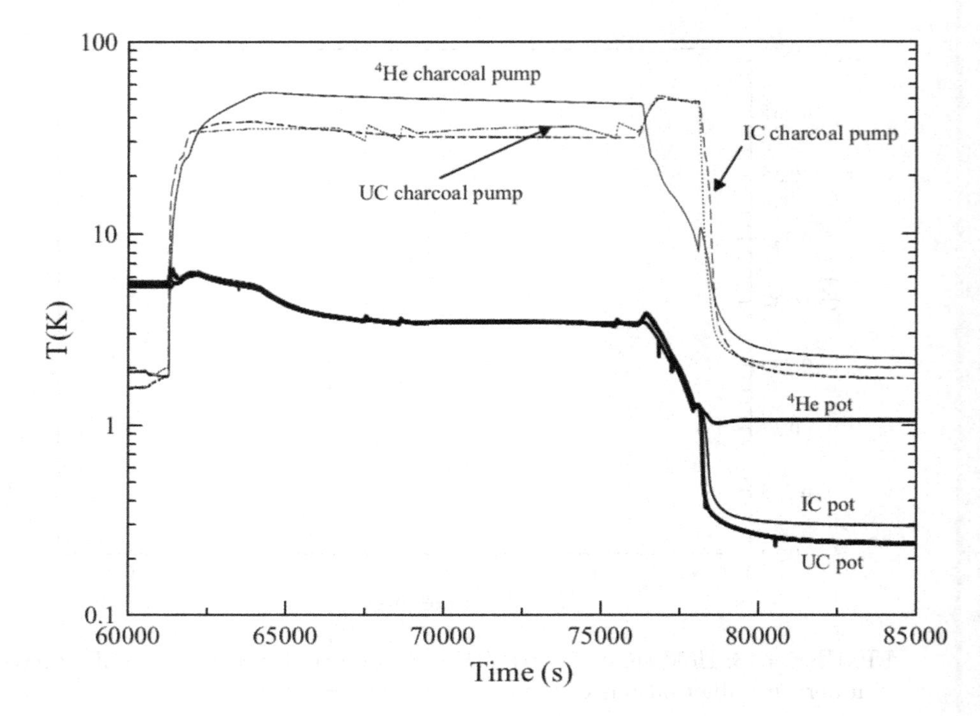

FIGURE 4.15: Typical charcoal pumps heating cycle to bring the three evaporator pots at their operating temperatures in a 220 mK triple-stage refrigerator

final stage, at least. Of course, as for the previous system, it can also be used to provide a 1 K stage to other components.

Operating such a system needs to be carefully optimized in order to achieve the lowest temperature. In particular, both the ^{3}He pumps should be heated at the same time. In this way, the ^{3}He in the intermediate stage can be used to pre-cool the ^{3}He in the final stage while it is cooling down. This is thermally efficient because, as for the ^{4}He, the latent heat of ^{3}He decreases with temperature. The ^{4}He stage can be operated in the same way described for the ^{3}He/^{4}He system. An example of the full system cycle is shown in Fig. 4.15.

^{3}He/^{3}He/^{4}He systems can typically reach a temperature of 220 mK with a cooling power of some µW. In order to reach such a temperature, the vacuum created by the cryopump is similar to the best vacuum in the ^{4}He sorption coolers. This means that this system is pump limited as for the ^{4}He. Consequently, there is not an advantage to add another ^{3}He sorption cooler, as was done for the ^{3}He/^{4}He.

4.6 CONTINUOUS SYSTEMS

All of the systems studied so far have many advantages in terms of size, ease of operation, no cold moving parts, no external gas handling system, etc. However, operations involving charcoal pumps have intrinsically the time limitation of the cold phase. In other words, their cold phase lasts as long as the coldest liquid is still evaporating. When all the liquid has changed phase and all the gas has been cryosorbed in the charcoal pump, the system needs to be regenerated. For many applications this fact does not represent a big limitation. Experiments that require changing samples at the coldest stage after a few hours of data taking, for example, will not need a particularly long duration of the cold cycle. Other experiments, in particular astrophysics or dark matter, where very long integration times are necessary, might suffer some decrease in efficiency if the system needs to be regenerated periodically. The regeneration duty cycle for closed-cycle, single-shot systems rarely surpasses 95%. Although such a figure might not seem too bad, it is important to note that certain astrophysics experiments attempting to make full maps of the sky might be affected by having to interrupt the observations generating dead time in the data stream.

Some very recent efforts have been made in order to make sorption coolers work in pairs to maintain a "switched stage" always cold. Two different methods have been investigated so far but it is not difficult to imagine other ways to achieve continuous operations with sealed closed-cycle refrigerators.

The continuous systems always have two parallel refrigerators capable of achieving the needed temperature and the needed cooling power. Two techniques will be described that are capable of alternatively connecting and isolating the two refrigerators while they are in the "cold" and "regeneration" phases.

4.6.1 Passive and active convective switched stage

We have already discussed the possibility of using a convective flow of helium gas to thermally connect or disconnect two surfaces at different temperatures. Convection uses gravity: as long as the cold surface is above the hot surface, convection is present and thermal conductivity is high. If, on the other hand, the hot surface is above the cold surface, then convection is inhibited and the thermal conductivity is low. Such a behaviour resembles the operation of a diode in an electrical circuit in that current flows only in one direction.

In Fig. 4.16 a schematic representation of the on and off states of a passive convective heat switch are shown.

A continuous $\sim$ 300 mK switched stage using a passive convective heat switch is shown in Fig. 4.17. This system uses two different ^{3}He fridges. In Fig. 4.17 only the two cold pots are shown. The two heat switches are connected to the two pots and they are also connected in parallel. The other side of both heat switches is connected to a common flange which is called *switched*

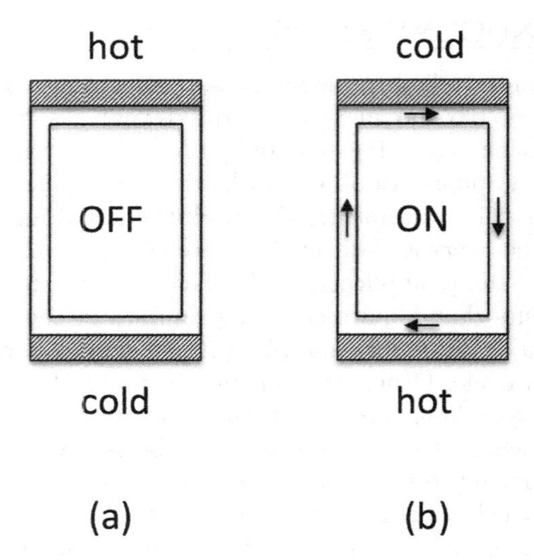

FIGURE 4.16: Passive convection switch in the presence of gravity. (a) No convection: thermal conductivity is dominated by the tubes. (b) Convection: thermal conductivity is dominated by convective motion of gas inside the system.

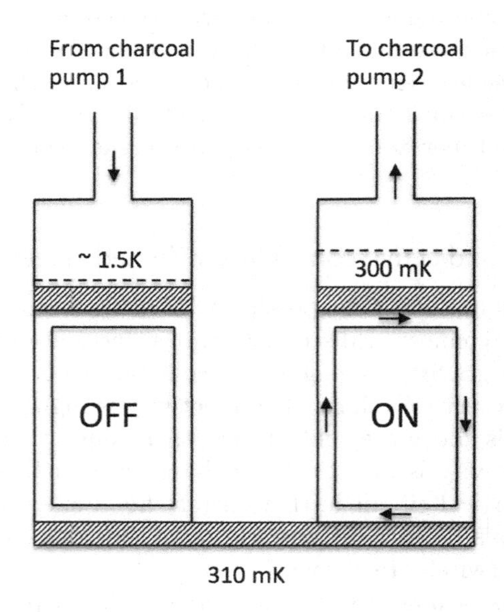

FIGURE 4.17: Continuous stage at 310 mK. Charcoal pump 2 is keeping the top of the convective switch colder than the switched stage thus allowing high thermal contact. Charcoal pump 1 is regenerating and convection is inhibited because the top of the switch is hotter than the bottom.

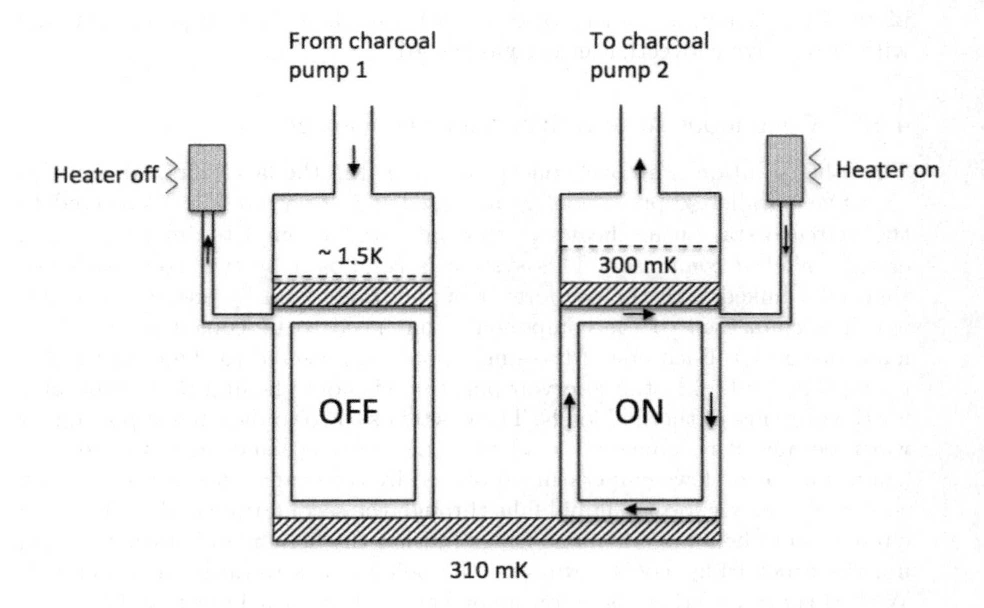

FIGURE 4.18: Continuous stage at 310 mK using two active convective heat switches

stage. When one of the sorption coolers is cold, the heat switch is activated and so the switched stage is connected to the coldest part of the fridge in order to reach the operative temperature. At the same time the other heat switch is off, thus allowing the other fridge to be regenerated. Note that in this system the switches do not have an off conductance exactly equal to zero. This means that when the non-operative fridge is regenerated, there will be a heat load through the switch that heats up the switched stage. In addition, this heat is not constant but it is dependent on the phase of the cycle (e.g., heat or cool the pump). As a result of this variable heat load through the tube, the temperature of the switched stage is subject to variations. To keep the temperature constant, a PID (proportional-integral-differential) control can be used. This helps to keep a constant temperature but it will need to be set to a slightly higher temperature than the lowest temperature achievable with a single sorption cooler. For example if the single sorption cooler has a temperature of 300 mK, the temperature of the switched stage can be higher than 50 mK when the same load is applied.

A slightly different variation of such a switched stage can use active convective heat switches so as to have major control of the thermal conductivity of the switch. This is achieved by controlling the temperature of the charcoal pumps by means of heaters and weak links to the cold reservoir (the base plate of a mechanical cooler, for example). Fig. 4.18 shows a possible design with two active convection heat switches [61].

4.6.2 Continuous stage with switched condenser

The other solution is a novel concept developed in the last years. Also in this case two parallel sorption coolers are used, but they are not connected to the switched stage using heat switches but using a complete different system called *switched condenser*. This system is composed by two pots, each one thermally linked with the evaporator of the fridge, and by one reservoir pot which is connected to the components that need to be cooled (so the heat loads are on it). Each one of the single pots is connected to the reservoir pot through two tubes. In the reservoir pot, there is liquid helium that evaporates for the presence of thermal loads. The gas tends to go to the coldest pot (one is warm because it is connected to a fridge that is cycled) and there it condenses again due to the low temperature provided by the evaporator of the sorption cooler. The newly formed liquid falls through the second tube and so there is a circulation of helium. When one fridge finishes the helium and starts to warm up, the other fridge cools down and the helium goes towards the other pot. With this system, there is a continuous circulation of helium that keeps the reservoir pot at the low temperature. In reality, there is a small heat coming through the tubes from the warm pot, which varies its temperature following the temperature of the associated evaporator. This means that also in this case the switched stage (so the reservoir pot) has a temperature fluctuation and the easiest way to solve it is to use a PID as in the previous case. A similar fridge is presented in Klemencic et al. [43].

While the first system is based on heat switches, so on conduction or convection of the helium gas or liquid, the second system is based on the latent heat of the liquid in the reservoir pot. In the second system, the parasitic heat load comes through the tubes from the warm pot to the reservoir pot, so it can be reduced by making the tubes longer. While, in the first case, the parasitic heat load is due to a non-zero off conductance of the switch that can be composed both by the conduction through the components of the switch and by the remaining gas in the switch that is not completely adsorbed by the pump.

4.7 DESIGN PROCESS FOR A SINGLE-STAGE ^{3}HE COOLER

Design of a ^{3}He sorption cooler requires consideration of all of the theory discussed in the previous sections.

In order to design any cooler, one must first identify the constraints. The most important one is the operational temperature of the sorption cooler. Then, applying other constraints during the designing phase is dependent on the experiment where a sorption cooler needs to be deployed. For example, some experiments have a limited space available so there are mechanical constraints to consider in designing the system. Moreover, in other situations it is important to set hold time and heat load.

Generally, a single stage ^{3}He sorption cooler is designed to have more than one tube between the condenser and the evaporator and one bigger tube from the evaporator to the pump. The idea of having more than one tube between the condenser and the evaporator is due to the fact that, this way, the liquid is cooled by the convection of the gas in the tubes during the cooling to reach the condenser temperature. This reduces the condensation time and so the cycling.

4.7.1 Mechanical constraints

If the fridge needs to be designed for an experiment with some mechanical constraints, it is possible to design it as follows. First of all, we need to set all the dimensions of the sorption cooler, this means all the diameters and lengths of tubes and the diameters and heights of the evaporator and the pump. Once these constraints are set, it is possible to run a simulation to find the best sorption coolers desired. The idea is to create several combinations of helium quantities and load and then verify that each combination generated respects Eq. 4.25 and that the pump is able to adsorb all the helium. In order to respect the last criteria, it is possible to look at Fig. 3.3 and check which is the quantity of helium adsorbed by the charcoal. The quantity of charcoal is set by the internal volume available on the cryopump. In particular, it is possible to compute the grams of charcoal available considering an averaging density of the charcoal of 0.4 g/cm^3. If both criteria are respected, then the combination generated by the simulation is acceptable and can be chosen. A flowchart of the process used to design the fridge is presented in Fig. 4.19.

In Fig. 4.20, a plot is presented where hold time is presented as a function of charging gas and heat load applied. In this particular case the dimensions chosen for the fridge are presented in Table 4.2. In the table, the subscript EC means from the evaporator to the condenser, while the subscript CP means from the condenser to the pump. E and P refer to evaporator and pump, respectively. ID refers to the internal diameter. The thickness of the tubes from the evaporator to the condenser is 0.5 mm. For the results presented in Fig. 4.20, the presence of two tubes from the evaporator to the condenser has been considered.

l_{EC}	ID_{EC}	l_{CP}	ID_{CP}	r_E	h_E	r_P	h_P
150.00	7.00	90.00	17.00	30.00	40.00	30.00	55.00

TABLE 4.2: Fridge parameters used to create Fig. 4.20. All dimensions are in mm.

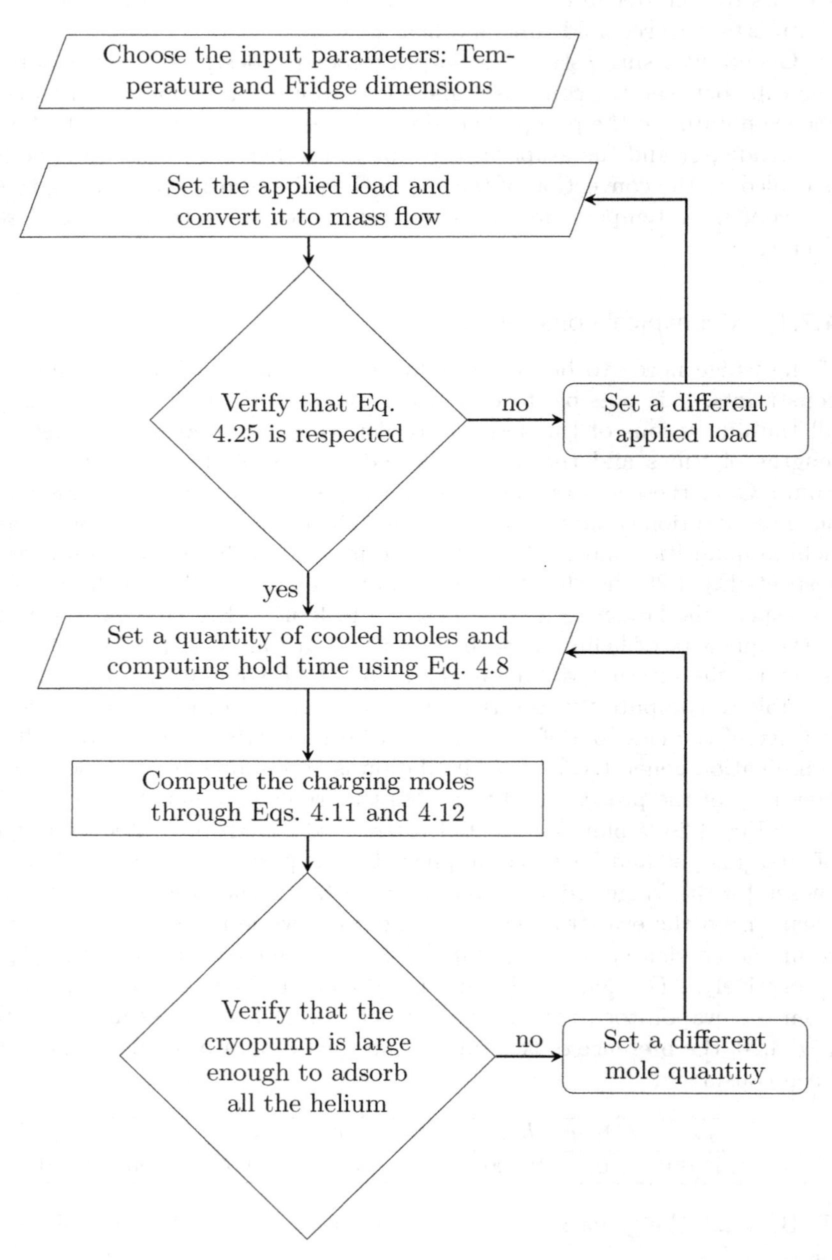

FIGURE 4.19: Flowchart of the process to design a single stage ^{3}He sorption cooler given mechanical constraints

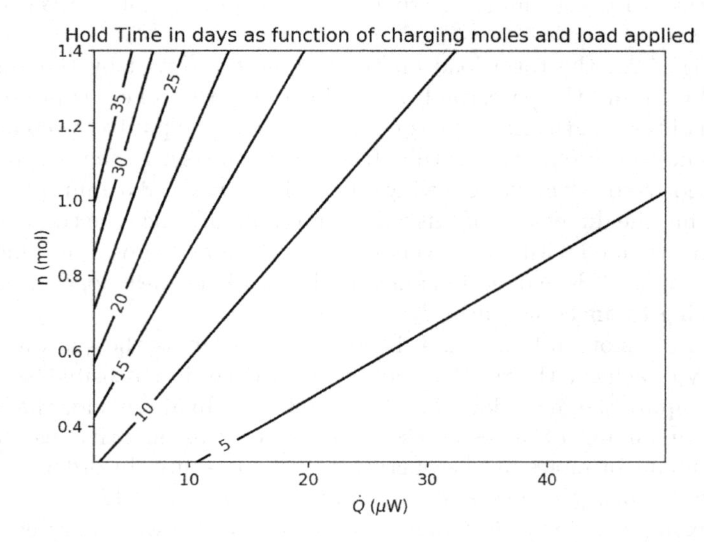

FIGURE 4.20: Hold time as a function of charging gas and heat load applied

4.7.2 Cryogenics constraints

In the case where the sorption cooler is constrained by heat load, the strategy to design it is different. In particular, it is possible to generate a simulation where only heat load and temperature are fixed. Once this is done, it is possible to generate an array of the number of moles in the liquid phase and consequently the hold time through Eq. 4.8.

Next, we must find the required dimensions of the different components. For the helium pot, by considering the molar volume, it is possible to compute the internal volume required to hold the liquid; from this volume, it is possible to generate a combination of radii and heights that satisfy this condition. Then for the tubes from the pot to the condenser and from the condenser to the pump, a series of radii and lengths are generated and then this combination needs to satisfy Eq. 4.24. Finally for the pump the internal volume needs to consider the quantity of charcoal necessary. This one is fixed by the quantity of helium that needs to be adsorbed and it is possible to estimate using Fig. 3.3. At the end of the simulations we will have a series of combinations of 8 geometrical parameters (lengths and radii of the tubes and radii and heights of the pot and the pump) and two cryogenic values: the quantity of gas necessary at the lowest temperature and the hold time. Between these sets of 10 parameters, it is possible to choose which is the best combination. Finally in order to compute the charging gas, it is possible to invert the self-cooling loss function and the condensation efficiency, allowing to compute the quantity of gas requested at room temperature. In Fig. 4.21, a flowchart is presented with the process to design and ^{3}He single-stage sorption cooler, given cryogenic constraints.

In Fig. 4.22, the total load on the evaporator (given by the sum of the external load and the parasitic heat load through the tubes) is presented as a function of the length and radius of the tubes from the pot to the condenser. In order to make these plots, the tube from the condenser to the pump has been considered fixed with the following dimensions: $r_{cp} = 8.5$ mm and $l_cp = 70$ mm. Fixing the dimension of this tube is not critical, since the condenser stage is usually provided with a relatively high cooling power (given by a mechanical cooler or by an ^{4}He bath). This means that a higher heat load on this stage is not going to affect significantly the stage itself.

Given the plots in Fig. 4.22, it is possible to compute the quantity of moles necessary to achieve the required hold time. Indeed, it is possible to apply Eq. 4.4 to compute the mass flow (in mol/s) and then divide this for the hold time to find the amount of moles necessary. However, this quantity just computed is the amount of moles at the operational temperature. In order to compute the charging gas, it is necessary to apply Eqs. 4.11 and 4.12.

An example code for designing a ^{3}He cooler in this way is presented in the Appendix in Section B.2.

4.8 DESIGN PROCESS FOR A DOUBLE-STAGE ^{4}HE-^{3}HE COOLER

In Section 4.7, we explained the process to design a ^{3}He single-stage fridge. However, the mechanical cooler is not always cold enough to liquefy the ^{3}He, so there is a necessity to add an intermediate stage to pre-cool the gas furthermore. This additional stage is usually a ^{4}He sorption cooler. This buffer stage not only allows the liquefaction of ^{3}He, but it also reduces the parasitic heat load on the ^{3}He pot.

The main difficulty in designing this fridge is related to the quantity of ^{4}He necessary in the buffer stage. Indeed, the ^{3}He stage can simply be designed as explained in Section 4.7 considering the condenser temperature of 1 K. Instead for the buffer stage, it is necessary to consider that the ^{4}He needs to provide not only the hold time requested with the additional load, but also the cooling power to cool the ^{3}He.

In order to compute the ^{4}He quantity, it is necessary to compute the amount of energy that needs to be removed from the ^{3}He. From Eq. 4.11, it is evident that not all the ^{3}He condenses at 1 K but only a quantity n_3. With this information it is possible to compute the energy that needs to be removed as:

$$E_1 = n_3 \int_{T_l}^{T_{mc}} C_3(T)dt + L_3(T_l)n_3 + n_3 \int_{1}^{T_l} C_3(T)dt \qquad (4.30)$$

where the first term is the cooling in the gas phase, the second is the energy at the liquefaction temperature and the last one is the cooling in the liquid phase down to 1 K. A crucial point in Eq. 4.30 is to determine the

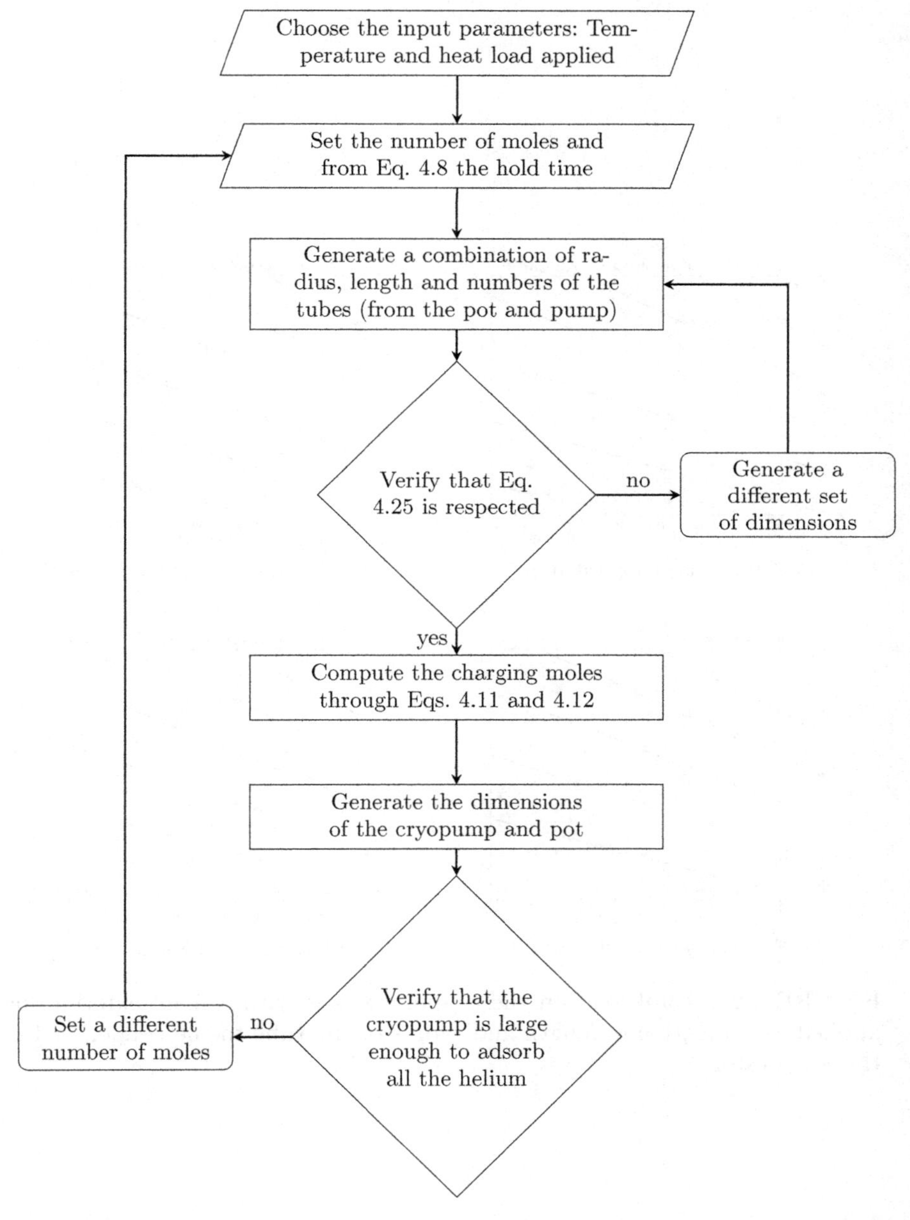

FIGURE 4.21: Flowchart of the process to design a single stage ^{3}He sorption cooler given cryogenics constraints

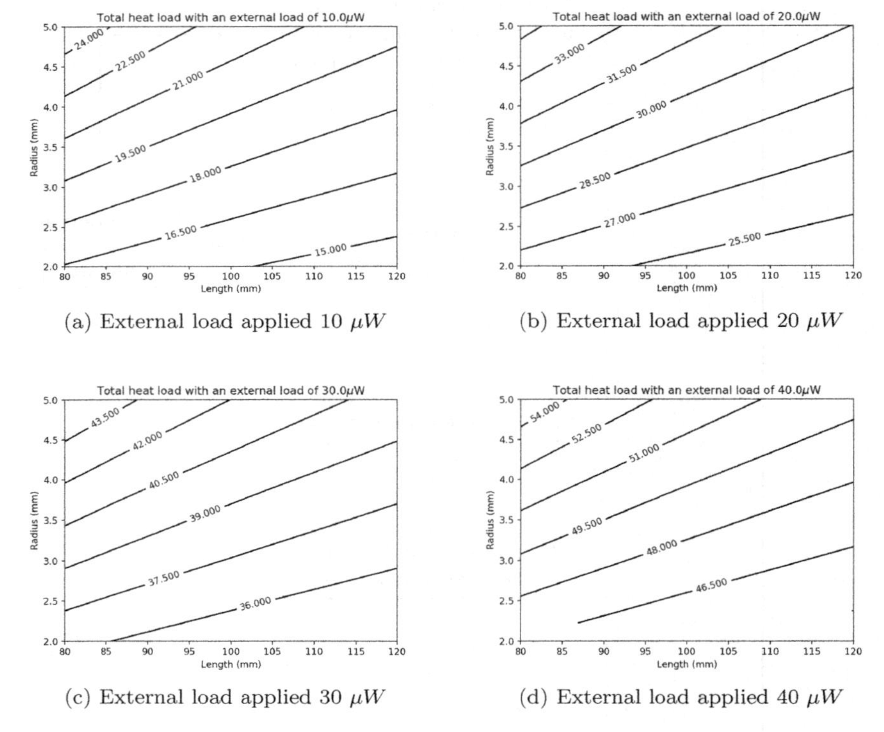

(a) External load applied 10 μW

(b) External load applied 20 μW

(c) External load applied 30 μW

(d) External load applied 40 μW

FIGURE 4.22: Total load on the evaporator pot with a fixed external load applied as a function of length and radius of the tube to the condenser from the evaporator

condensation temperature. In order to do that, it is necessary to compute the pressure at different temperatures using the gas perfect law:

$$PV = nRT \tag{4.31}$$

when the pressure is greater than the vapour pressure at the same temperature, that is the condensation temperature that can be considered.

In case the ^{4}He stage is added only to reduce the parasitic heat load through the tubes and the mechanical cooler is able to provide a temperature low enough to condense the ^{3}He, Eq. 4.30 becomes:

$$E_1 = n_3 \int_1^{T_{mc}} C_3(T)dt \tag{4.32}$$

The energy E_1 is only the energy requested to cool the ^{3}He. In order to maintain the parasitic heat load constant during the operation of the ^{3}He stage, it is necessary to consider the cooling energy that the ^{3}He stage is able to provide. To compute the cooling energy, it is important to notice that the quantity of ^{3}He is not anymore n_3 but it is n_3' which is the liquid after that has been pumped to reach its minimum temperature. To compute n_3', it is possible to use Eq. 4.12 that consider the self-cooling of the liquid. The self-cooling is not done at the expense of the ^{4}He, so during this short time the quantity of ^{4}He remains constant. When the ^{3}He reaches its minimum temperature, the cooling energy available is given by:

$$E_2 = n_3' L_3(T_{min3}) + \dot{Q}_{p3} t \tag{4.33}$$

where $\dot{Q}_{p3}$ is the parasitic heat load on the ^{3}He stage and t is the hold time.

Using Eqs. 4.30 (or 4.32) and 4.33, it is possible to compute the energy that the ^{4}He stage needs to provide and so the quantity of liquid necessary. In particular, the quantity is given by:

$$n_4' = \frac{E_1 + E_2 + \dot{Q}_{p4} t}{L_4(T_{min4})} \tag{4.34}$$

where n_4' is the final quantity of ^{4}He and $\dot{Q}_{p4}$ is the parasitic heat load on the ^{4}He stage. In order to compute the charging gas for the ^{4}He stage, it is necessary to reverse Eq. 4.12 to compute the liquid available before the pumping and Eq. 4.11 to include the condensation efficiency.

It can happen that the ^{4}He stage is used to cool not only the ^{3}He stage, but also some components of the experiment. In this case the quantity of ^{4}He requested is given by:

$$n_4' = \frac{E_1 + E_2 + (\dot{Q}_4 + \dot{Q}_{p4}) t}{L_4(T_{min4})} \tag{4.35}$$

where $\dot{Q}_4$ is the power applied on the ^{4}He stage.

compensation temperature [illegible] is necessary to compute the [illegible] at different [illegible]

$$[illegible] \qquad (4.31)$$

when the pressure [illegible] the [illegible] temperature, that is the [illegible]

[illegible] through the [illegible] to [illegible]

$$[illegible] \qquad (4.32)$$

The energy [illegible] is only the [illegible] to [illegible] the [illegible] stage, it is necessary [illegible] to [illegible] quantity of [illegible] been pumped [illegible] to [illegible] 4.12 that [illegible] at the [illegible] constant [illegible] cooling [illegible]

$$[illegible]$$

where [illegible] is the [illegible] time.

Using Eqs. [illegible] that the [illegible]

$$[illegible]$$

where [illegible] on the [illegible] it is necessary [illegible] pumping and Eq. [illegible]

It [illegible] happen that [illegible] also [illegible] required [illegible] given by

$$[illegible] \qquad (4.35)$$

where Q_R is the [illegible]

CHAPTER 5

Miniature Sorption Coolers - Part 2

DILUTION REFRIGERATION is the most used technique to reach temperatures below 100 mK. The typical dilution refrigerator system consists of a cryostat containing the cold stages serviced with external pumps, compressor and a gas handling system. Many commercial systems are available today capable of achieving remarkable cooling power at 100 mK, often in excess of 500 μW. These kinds of systems are designed for the standard laboratory use and tend to be relatively expensive. If the experiment does not require such high cooling powers, but requires ease of operation, limited external plumbing and low power dissipation, then miniature self-contained dilution refrigerator systems can be considered. After a brief and non-exhaustive introduction to the dilution refrigeration concepts, this chapter will show few different designs of small dilution systems. In this chapter, the reader will find all the relevant formulas and tables to be able to design its own custom miniature dilution refrigerator.

5.1 INTRODUCTION TO DILUTION REFRIGERATION THEORY

The sorption coolers described in the previous chapter, with some elaborate design, can reach a temperature of 220 mK. However, for some particular experiment, there can be the necessity to reach a temperature below 100 mK. *Dilution refrigerators*[1] exploit the enthalpy difference between pure ^{3}He and diluted ^{3}He into ^{4}He and can reach temperatures as low as $\sim$ 2 mK. In order

[1]ADR (Adiabatic Demagnetization Refrigerators) can allow to reach similar temperatures or lower. However, this category of refrigerators is based on relatively strong magnetic fields that can interfere with the electronic if they are not properly shielded. For this reason, dilution refrigerators are usually the choice for sub-100 mK temperature range because they require no magnetic field and have much larger cooling capacity.

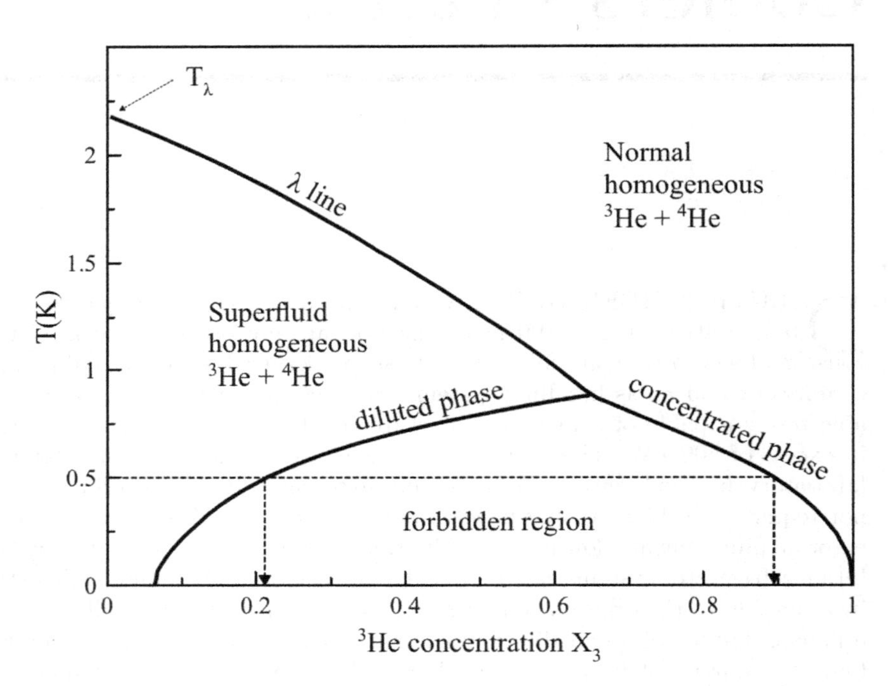

FIGURE 5.1: Phase diagram of an ^{3}He and ^{4}He Mixtures. This plot gives the ^{3}He concentration in the diluted and concentrated phases when below T = 0.85 K. For example, at T = 0.5 K, the ^{3}He concentration in the diluted phase is slightly above 20%, while in the concentrated phase it is about 90%. Above T = 0.85 K, the λ line divides the superfluid from the normal fluid phase. Note that at T = 0 K, there is pure ^{3}He in the concentrated phase and 6.6% ^{3}He and 93.4% ^{4}He in the diluted phase.

to understand the functioning of a dilution refrigerator, we need to understand the physics of mixtures of ^{3}He and ^{4}He at low temperatures.

5.1.1 Physics of ^{3}He and ^{4}He mixtures

The dilution refrigeration process exploits some remarkable properties of the liquid mixture of ^{4}He with its isotope ^{3}He. Above $T = 0.85$ K the two liquids are mixed homogeneously and ^{3}He can dilute into ^{4}He in any fraction. Below $T = 0.85$ K phase separation occurs and the liquids, if subject to gravity, will stratify with concentrated ^{3}He above and diluted ^{3}He into ^{4}He below. Fig. 5.1 shows the so-called λ-plot, i.e., the phase diagram of a mixture of ^{3}He and ^{4}He, where $x_3 = n_3/(n_3 + n_4)$ is the fraction of ^{3}He and $n_{3,4}$ are the number of atoms of, respectively, ^{3}He and ^{4}He. Knowing x_3 immediately gives $x_4 = n_4/(n_3 + n_4)$.

In a ^{3}He and ^{4}He mixture, the temperature of the superfluid transition of ^{4}He is dependent on the concentration of ^{3}He and it varies as [35]:

$$T_\lambda \propto (1 - x_3)^{2/3} \tag{5.1}$$

Above ~ 2.17 K, the mixture behaves like a classical fluid. With decreasing the temperature, quantum effects become more and more important. Liquid ^{4}He, with its integer spin, is subject to Bose-Einstein condensation and becomes superfluid. Liquid ^{3}He, on the other hand, with its half-integer spin is subject to Fermi-Dirac statistics and thus subject to Pauli exclusion principle. This prevents ^{3}He to become superfluid until $T = 0.94$ mK when the energy is low enough for two atoms to pair up and form a boson state. In the normal range of temperatures of dilution refrigerators, only ^{4}He becomes superfluid while ^{3}He stays normal. Being fermions, ^{3}He atoms in the diluted phase will fill the lowest energy states available in anti-parallel spin pairs. Additional atoms then must occupy states at increasingly higher energies. In the theory of Fermi gas, assuming N *non-interacting* particles confined in a volume V, at temperature $T = 0$ K, the Fermi energy[2] is defined as the energy of the highest occupied quantum level and it is given by:

$$E_F = k_B T_F = \frac{\hbar^2}{2m}\left(\frac{3\pi^2 N}{V}\right)^{2/3} \tag{5.2}$$

If we want to write the same equation for the ^{3}He atoms, we need to take into account that there is a *weak* interaction among the ^{3}He atoms with the ^{4}He in the diluted phase or with themselves in the concentrated phase. It turns out that we can apply Eq. 5.2 providing that we use an effective mass m^* instead of m. In case of a diluted phase with concentration $x_3 = 6.6\%$, which we will see shortly that it represents an important concentration, the

[2]Not to be confused with Fermi level, which is defined for $T > 0$ and is the total chemical potential, i.e., the *thermodynamic work* needed to add one fermion from infinity to the body. Sometimes, especially in semiconductor physics, the two terms are used in the same context.

effective mass is $m_3^* = 2.45m_3$, where m_3 is the mass of an ^{3}He atom. We have:

$$k_B T_F(x_{3D}) = \frac{\hbar^2}{2m_3^*}\left(\frac{3\pi^2 x_{3D}}{V}\right)^{2/3} \tag{5.3}$$

where x_{3D} is the concentration of ^{3}He in the diluted phase and V is the volume. Eq. 5.3 then justifies the proportionality shown in Eq. 5.1.

The *Landau-Pomeranchuk Theory* gives a good description of the concentrations of the dilute phase[3]. According to the Landau-Pomeranchuk theory, the ^{3}He and ^{4}He concentrations in the diluted phase can be written as:

$$x_4 \simeq 0.85T^{3/2}e^{-0.56/T} \tag{5.4}$$

$$x_3 = 0.066\left(1 + 8.3T^2 + 9.4T^3\right) \tag{5.5}$$

As is evident from Eq. 5.5 and from Fig. 5.1, the concentration of ^{3}He reaches the value of 6.6% when $T = 0$ and this is one of the key elements of the dilution refrigeration.

5.1.2 Principle of operation of a dilution refrigerator

At temperatures below 0.1 K, the highest concentration possible in the dilute phase is 0.066 (6.6%). As a result, if the mixture at the start of the process has a higher concentration than this and is cooled to around 100 mK, then it will always end up with a dilute phase with $x_{3,D} = 0.066$ and a concentrated phase of pure ^{3}He ($x_{3,C} = 1$). It is important to note that even as T approaches 0 K, the solubility limit is still 0.066.

Let us consider a pot containing ^{3}He plus ^{4}He at a temperature $T <$ 0.1K (see Fig. 5.2). Let us consider a non-physical situation where the top lighter layer contains pure ^{3}He and the bottom heavier layer contains pure superfluid ^{4}He. We have seen already that the lighter ^{3}He atom has a larger zero point fluctuation amplitude. This means that the ^{3}He atom occupies a large volume than the ^{4}He atom. In the pure phase, the ^{3}He atoms have an average distance d_1 which is larger than the average distance between ^{4}He atoms in the bottom layer. This is because the ^{4}He, being heavier has a smaller zero point fluctuation amplitude therefore occupying a smaller volume. The binding energy of two ^{3}He atoms in the pure phase is smaller than the binding energy of the same ^{3}He atom once in the pure ^{4}He phase because the distance ^{3}He - ^{4}He is less than the distance ^{3}He - ^{3}He. Therefore a ^{3}He atom is energetically favoured to cross into the bottom phase because it will be subject to a stronger binding energy. We will see shortly that the crossing of an ^{3}He atom through the phase boundary from the concentrated to the diluted phase requires heat to be absorbed, i.e., cooling is produced. Once the ^{3}He atom has

[3] The concentrated phase is practically pure ^{3}He.

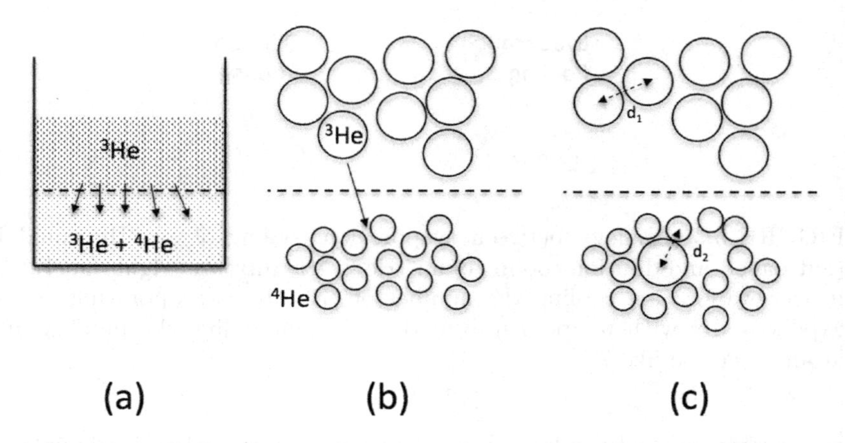

FIGURE 5.2: Principle of dilution cooling. In (a) we have a pot in an idealized non-physical situation with pure ^{3}He floating above pure ^{4}He. ^{3}He atoms will diffuse into pure ^{4}He. Each ^{3}He atom crossing the phase separation will subtract a small amount of heat from the wall of the container equal to the enthalpy difference between concentrated and diluted phase. (b) and (c) ^{3}He atoms being lighter than ^{4}He atoms, will have larger zero point fluctuations, i.e., occupy a larger volume. The average distance d_1 of two ^{3}He atoms is larger than the average distance d_2 between an ^{3}He and an ^{4}He atom. Since the binding energy will be stronger for shorter distances, the ^{3}He atoms prefer to diffuse into ^{4}He. Once diffused in the other phase, ^{3}He atoms, being fermions, will occupy all the energetic levels available up to the Fermi level until a concentration of 6.6% is reached. At this point, the crossing rate of ^{3}He atoms decreases to zero and the dilution stops.

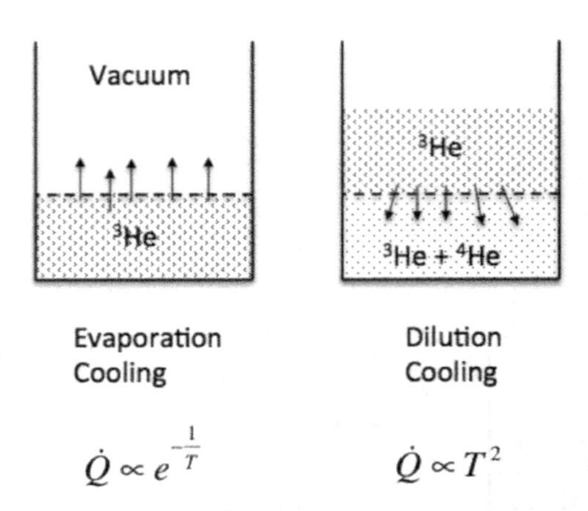

FIGURE 5.3: Analogy between evaporation cooling of an ^{3}He liquid bath (left panel) and dilution cooling of an ^{3}He + ^{4}He mixture (right panel). While in the evaporation cooling the number of ^{3}He atoms evaporating decreases exponentially with temperature, in the dilution cooling the number of ^{3}He atoms decrease like T^2.

crossed the phase boundary it is a fermion subject to Pauli principle. This means that, as other ^{3}He atoms cross, they pair with anti-parallel spin and occupy the lowest available energy level until they reach a concentration of ^{3}He which will make the crossing of additional ^{3}He atoms not energetically convenient. The concentration at $T < 100$ mK is practically 6.6%. The bottom superfluid phase is a Bose-Einstein condensate of ^{4}He atoms. If its temperature is < 500 mK, its entropy, viscosity and specific heat are very small tending to zero. The superfluid ^{4}He phase acts as a sort of "vacuum" where the ^{3}He atoms can move almost freely like a gas (see Fig. 5.3). Once the concentration of minimum energy has been reached, in order to continue the cooling process we need to force other ^{3}He atoms to cross the phase boundary. If we remove ^{3}He atoms from the diluted phase, then we make energy levels available for other ^{3}He atoms so they can cross and keep cooling. We see therefore that the *^{3}He circulation rate* is the most important parameter when calculating the cooling power of the dilution refrigerator.

Another way to describe the dilution is by noticing that decreasing the concentration of ^{3}He in the dilute phase moves the two phases away from equilibrium. As such, a ^{3}He dilution flow will occur from the concentrated to the dilute phase as the two do not interact at this temperature and the ^{3}He experiences effectively a mechanical vacuum [46]. As the enthalpy of dilute ^{3}He is higher than that of concentrated ^{3}He, this dilution is an endothermic process, i.e., absorbs heat from the surrounding.

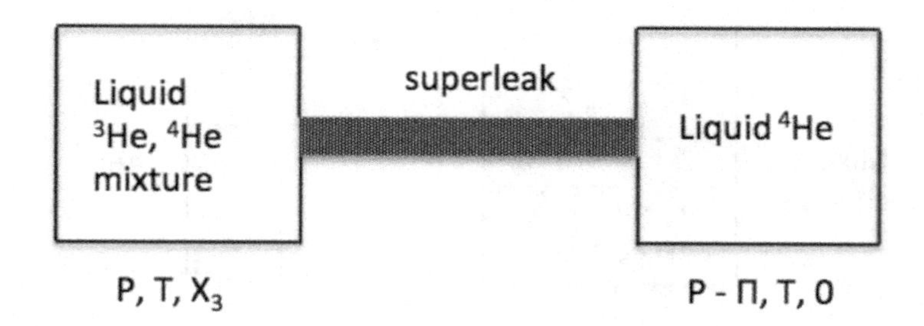

FIGURE 5.4: Two vessels connected with a superleak. Osmotic pressure Π develops between the two vessels. The mixing chamber and the still of a dilution refrigerator can be considered as two vessels connected by a non-ideal superleak, i.e., ^{3}He molecules can still pass through.

We now understand that, in order to produce cooling, ^{3}He must be continually removed from the dilute phase.

In order to explain the circulation of ^{3}He, we introduce the concept of *osmotic pressure*. To understand how this pressure acts, we can refer to Fig. 5.4. Here we follow the argument as presented in Pobell [66].

Here, we have two vessels connected by a superleak. In the first vessel, we have a mixture of the two He isotopes. In the second one we have pure ^{4}He. In this situation we can use Van't Hoff's equation. Van 't Hoff's equation states that a substance dissolved in a fluid medium behaves as if it were in a vacuum, and so exerts on the walls of the containing vessel a pressure which is precisely that which it would exert were the solvent imagined removed and the dissolved substance imagined present in a gaseous form. If we have a low concentration of ^{3}He in ^{4}He, then we can assume that the ^{3}He behaves as an ideal gas. Van 't Hoff's equation then becomes [66]:

$$\Pi V_{m,4} \simeq xRT \tag{5.6}$$

where $V_{m,4}$ is the molar volume of ^{4}He. Now we consider the configuration of the cold stage of a dilution refrigerator where the still is connected with a capillary tube to the diluted liquid inside the mixing chamber (see Fig. 5.5). If we apply Eq. 5.6 to such a configuration we have:

$$\Pi_{mc} - \Pi_{still} \simeq \frac{(x_{mc}T_{mc} - x_{still}T_{still})R}{V_{m,4}} \tag{5.7}$$

where the subscript mc stands for mixing chamber. If no ^{3}He is removed from the still, then the two chambers have exactly the same osmotic pressure. For example, if the mixing chamber is at $T = 10$ mK, then we know that the concentration in it will be very close to 6.6%. If the still is at its typical temperature $T = 0.7$ K, then Eq. 5.6 gives:

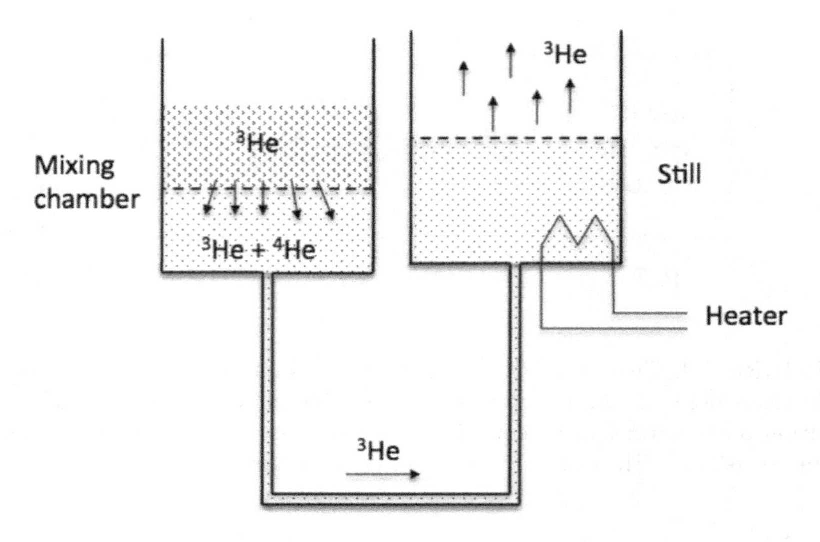

FIGURE 5.5: Mixing chamber connected to the still. Osmotic pressure Π pushes ^{3}He to the still. The still is heated to a temperature where practically only ^{3}He atoms evaporate ($T \sim 0.7$ K). Evaporation of ^{3}He atoms depletes the diluted phase in the mixing chamber making states available for ^{3}He atoms to cross the phase separation from the concentrated to the diluted phase thus generating cooling.

$$x_{still} = x_{mc} \frac{T_{mc}}{T_{still}} \simeq 0.1\% \tag{5.8}$$

As soon as we unbalance the relative concentrations between the 6.6% of the mixing chamber and the 0.1% of the still, for example by pumping away the ^{3}He evaporating from the still, an osmotic pressure difference will develop. In particular, if we reduce the concentration of the still, then the osmotic pressure in the mixing chamber will be higher than the osmotic pressure in the still and ^{3}He will be drawn from the concentrated phase toward the diluted phase in the mixing chamber. The maximum osmotic pressure develops if we set to zero the concentration in the still. We have:

$$\Delta\Pi_{max} = \frac{x_{mc} T_{mc} R}{V_{m,4}} \simeq 20mB \tag{5.9}$$

A more rigorous treatment that considers the non-ideality of Helium will give a temperature independent osmotic pressure, for $T < T_F/3$ numerically similar to the above result. The osmotic pressure is responsible for the flow rate of ^{3}He from the mixing chamber to the still and ultimately drives the cooling in a dilution refrigerator.

We already mentioned that the osmotic pressure develops when we connect the mixing chamber, i.e., the chamber where the ^{3}He plus ^{4}He mixture is separated into two phases, to a separate pot (still) connected to the diluted phase in the mixing chamber. If we manage to extract ^{3}He atoms from the diluted phase we make states available for additional ^{3}He atoms to cross the phase separation and generate cooling. If the still is heated to a temperature such that ^{3}He evaporates (and not ^{4}He), then for every ^{3}He atom evaporated, one ^{3}He atom can cross the phase separation and produce cooling. Again, it is evident that the cooling is more effective the more ^{3}He atoms cross the phase separation boundary. The process described will continue until all the ^{3}He in the mixing chamber is exhausted. If we want to achieve continuous cooling, then we need to re-condense the ^{3}He evaporated from the still. In standard dilution refrigerators, this is achieved by circulating the ^{3}He gas outside the cryostat through a gas manifold. The ^{3}He, after being purified is then re-injected in the system, pre-cooled and delivered to the concentrated phase in the mixing chamber. Effective pre-cooling of the returning ^{3}He is of paramount importance because it constitutes a spurious heat leak to the mixing chamber that can easily overcome the cooling power of the dilution process. We will see later that pre-cooling of the returning ^{3}He with heat exchangers is one of the most critical processes in designing dilution refrigerators.

Let us now calculate the cooling power of the dilution refrigeration process described. We have seen that the mixing chamber contains the two phases, as shown in Fig. 5.6, for which ^{3}He is continuously removed from the dilute phase and injected into the concentrate.

Since no external work is done in this process, the first law of thermodynamics shows that the cooling power at the mixing chamber is given by the difference in enthalpy of the ^{3}He crossing from the concentrated to the diluted phase:

$$\dot{Q} = \Delta H \dot{n}_3. \tag{5.10}$$

We need to estimate the enthalpies of concentrated and diluted phases. The crossing of ^{3}He atoms is a change of phase for which we must have:

$$\mu_D = \mu_C \tag{5.11}$$

since $\mu = H - TS$, we have:

$$H_D - TS_D = H_C - TS_C \tag{5.12}$$

where $H_{D,C}$ and $S_{D,S}$ are enthalpies and entropies per mole in the respective phases. The enthalpy change due to the mixing is then:

$$\Delta H = H_D - H_C = T(S_D - S_C) \tag{5.13}$$

we now need to estimate the entropies S_C and S_D in the two phases. Using the first law of thermodynamics $dU = \delta Q + dW$, we notice that during the

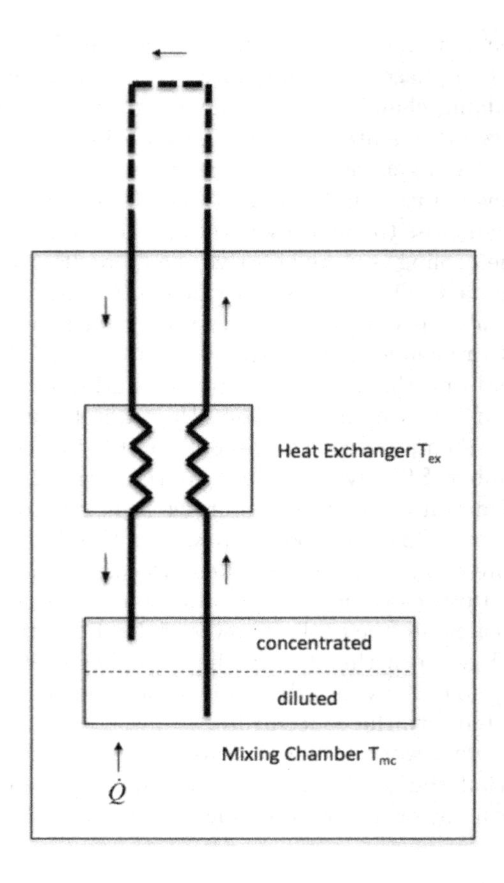

FIGURE 5.6: Diagram of mixing chamber

phase change in a liquid the volume change is negligible and therefore we can assume $dU = \delta Q$. We then can express the entropy:

$$S = \int \frac{\delta Q}{T} dT = \int \frac{dU}{T} dT = \int \frac{C(T)}{T} dT \tag{5.14}$$

Unfortunately, the data on specific heat of pure ^{3}He or on ^{3}He and ^{4}He mixture, with varying ^{3}He concentrations, is scarce (see Fig. 5.7).

In Fig. 5.7 we see some measurements of specific heat for pure and diluted ^{3}He [3]. We can see that a good fit to the data for $T \ll T_F$ can be obtained if we consider the ^{3}He atoms as a Fermi gas of quasi-particles weakly interacting with effective mass m^*. We are therefore justified to use the theory of a Fermi gas to get an analytic expression for the specific heat of the dilute mixture per mole [66]:

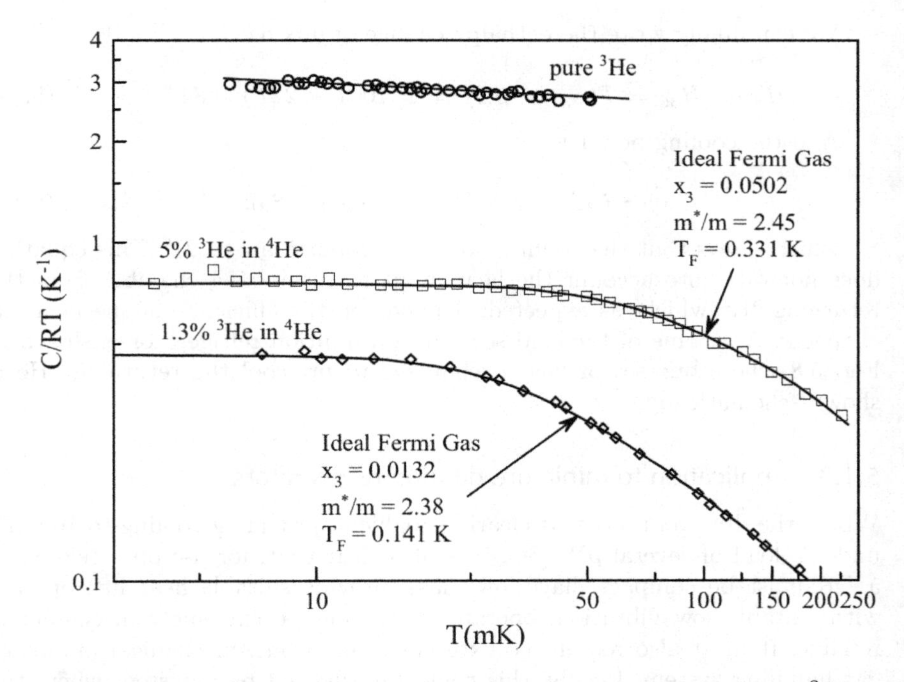

FIGURE 5.7: Measured ratio of specific heat per mole to RT of pure ^{3}He and mixtures ^{3}He/^{4}He. The solid lines are heat capacities of Fermi-Dirac particles with the parameters given in the figure (data from [3]).

$$C_{3D} = \frac{\pi^2}{2}\frac{T}{T_F}R \tag{5.15}$$

From Fig. 5.7 we also see that for $T \leq 40$ mK, the specific heat for pure ^{3}He is $C_3 = 2.7RT$. We can write now the specific heats for concentrated and diluted phases. Using Eq. 5.14:

$$C_{3D} = 106T = S_{3D} \tag{5.16}$$

$$C_{3C} = 22T = S_{3C} \tag{5.17}$$

We can finally write the enthalpy change of mixing:

$$H_{3D} - H_{3C} = T(S_{3D} - S_{3C}) = T(106T - 22T) = 84T^2 \tag{5.18}$$

And the cooling power is

$$\dot{Q} = \dot{n}_3\Delta H = \dot{n}_3(H_{3D} - H_{3C}) = 84\dot{n}_3T^2 \tag{5.19}$$

which shows that the cooling power is proportional to T^2. This equation does not take into account the heat input to the mixing chamber from the returning ^{3}He which, as expected, depends on the efficiency of the heat exchangers. A scheme of the cold section of a dilution refrigerator is shown in Fig. 5.8 where one set of heat exchangers to pre-cool the returning ^{3}He is shown schematically.

5.1.3 Application to miniature dilution refrigerators

Whilst the dilution process is clearly capable of providing cooling to 100 mK under a load of several μW [66], it is also clear that, for use on a telescope, a DR must be compact, have low mass, provide suitable heat lift, operate with suitably low vibration, operate in tilt and be extremely mechanically reliable. It must also require no external connections, mechanical pumps or gas handling system. Ideally, this could be achieved by a system where the ^{3}He is circulated within the cold stage.

Critically, the ability to operate away from the vertical is required for use on a focal plane as the telescope tracks across the sky. A tiltable miniaturized system is able to meet the temperature and heat lift requirements for bolometer arrays in CMB experiments, given a suitable flow rate and heat exchanger design.

5.2 DESIGN OF A CONDENSATION PUMPED 100 MK SYSTEM

A conventional dilution refrigerator circulates the ^{3}He gas outside the cryostat by means of powerful pumps. This allows the system to achieve relatively large cooling powers: modern dilution systems can have cooling powers at 100 mK in excess of 500 μW. This is achieved at the expense of bulky room temperature

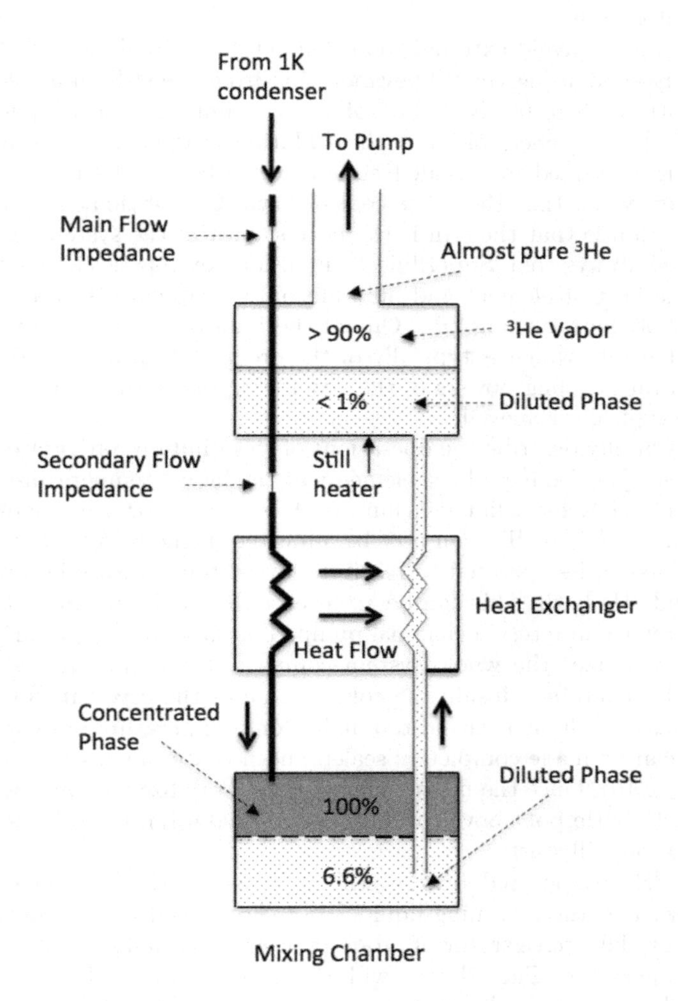

FIGURE 5.8: Scheme of a dilution refrigerator, figure based on London et al. [51]

gas handling system with large pumps and cumbersome large diameter pipes. If the requirement of large cooling power is somewhat relaxed, for example in small laboratory test systems where only cooling powers in the range of 1 and 10 μW are needed, then alternative solutions to conventional dilution refrigerator exist.

One way to avoid external room temperature circulation of ^{3}He gas consists in re-condensing the ^{3}He evaporating from the still on a cold surface at temperature 0.3 to 0.4 K [24] obtained by pumping over a liquid ^{3}He bath. Fig. 5.9 show a schematic of such a dilution refrigerator. The mixture ^{3}He + ^{4}He is contained in a sealed circuit and, when cold enough, is collected by gravity when the ^{3}He pot is cooled down. One obvious advantage of this configuration is that the still is the hottest part of the system and the gas is circulated always cold. Superfluid helium, for example, is confined in the still being the hottest element and, in addition, no capillary is needed to restrict the flow of the returning ^{3}He. One of the limitations is in the practical ^{3}He circulation rate which is typically of the order of 10 μmols^{-1}. With this circulation rate cooling powers of the order of no more than 10 μWK^{-1} at 100 mK are typically achieved.

Let's briefly describe the operations of this dilution refrigerator. The cycle starts with pre-cooling all the elements at the lowest temperature of the main reservoir (4.2 K for a liquid helium bath or $\sim$ 4 K for a mechanical cooler). The mixture ^{3}He + ^{4}He can now be admitted in the refrigerator and the top ^{3}He fridge can be operated to reach its lowest temperature by pumping over the liquid ^{3}He bath. This can be achieved either with an external mechanical pump or with an internal charcoal pump. This last option is clearly attractive because it makes the whole system compact. If the mixture of ^{3}He + ^{4}He gas is also contained inside the cold volume of the cryostat, for example in an expansion volume to avoid too high room temperature pressures, then the system can be made completely sealed and no external gas lines are required (see Fig. 5.10). Once the mixing chamber has stabilized at the lowest temperature of the ^{3}He pot above (usually around 400 mK), the still can be heated and dilution will start.

The ^{3}He evaporated by the still at T $\sim$ 0.7 K is re-condensed in the condenser and the returning liquid ^{3}He is pre-cooled by the continuous heat exchanger. This refrigerator, if properly designed, usually reaches $\sim$ 50 mK as base temperature. The dilution will continue as long as ^{3}He is re-condensed. The cycle duration is therefore limited by the amount of ^{3}He gas in the ^{3}He pot. Once the ^{3}He is exhausted, the cryopump needs to be regenerated to have another dilution cycle. See Fig. 5.11 for an example of a cooling cycle [84]. The design of this refrigerator contains a few critical elements that need to be properly calculated to ensure the maximum efficiency of the system. The top ^{3}He refrigerator needs to be powerful enough to ensure a temperature less than 400 mK to re-condense efficiently the ^{3}He. It also needs to contain enough ^{3}He to allow the refrigerator to run for the desired amount of time.

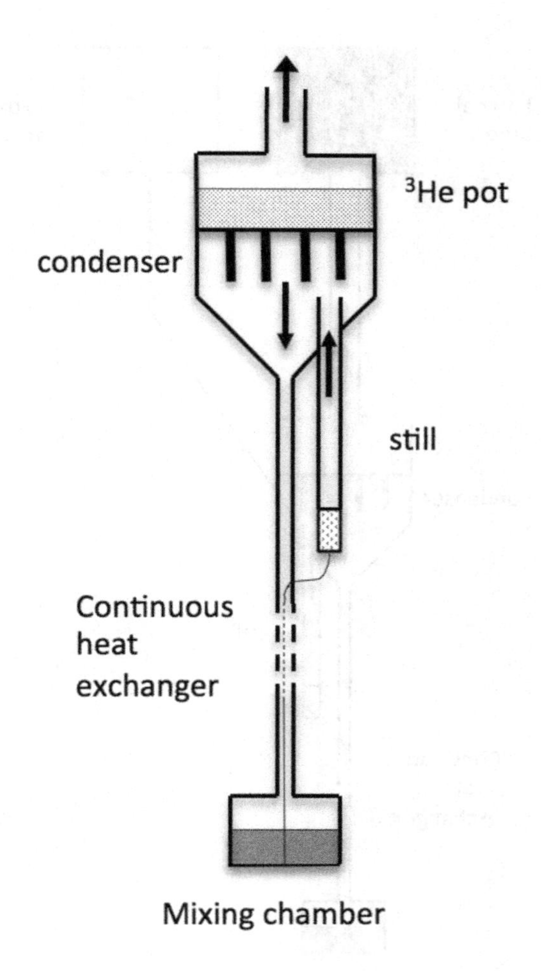

FIGURE 5.9: Edel'man dilution refrigerator with condensation pumping [24]

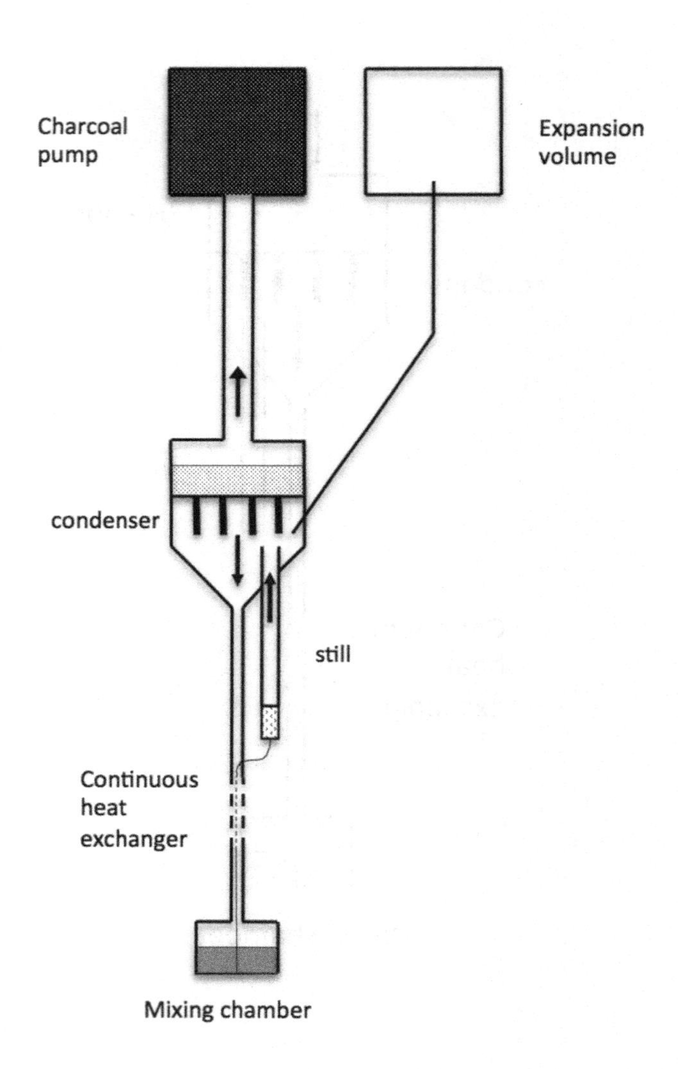

FIGURE 5.10: Sealed version of Edel'man dilution refrigerator

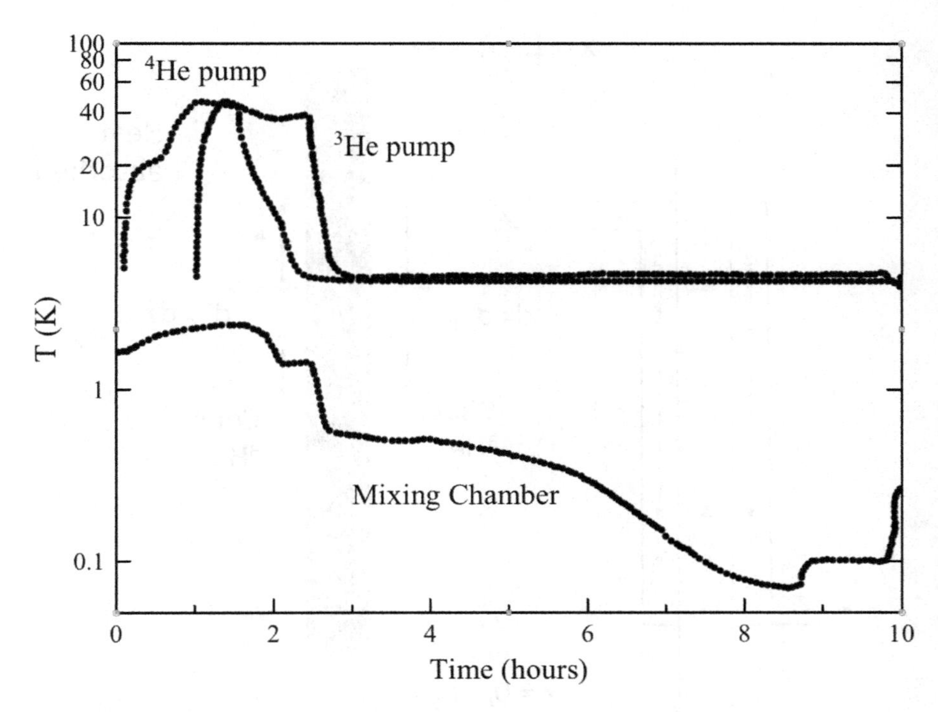

FIGURE 5.11: Cooldown of an Edel'man-like condensation pumped dilution refrigerator. The ^{4}He pump operates a 1 K pre-cooling fridge to the main ^{3}He fridge cooling the re-condensing plate for the ^{3}He gas circulating in the dilution system.

Later in the book, we will discuss a continuous system using a tandem of ^{3}He refrigerator.

The Kapitza resistance between liquid helium and copper in the top ^{3}He refrigerator, in the condenser and in the mixing chamber needs to be evaluated and taken into account. The continuous heat exchanger also needs to be properly designed because it will determine the temperature of the returning ^{3}He in the mixing chamber which represents one of the important heat inputs limiting the performances.

5.2.1 The continuous heat exchanger

We have seen in Section 2.1.4 a definition of Kapitza resistance as $R_K = \Delta T/\dot{Q}$, where ΔT is the temperature step generated by the applied power $\dot{Q}$. We have also seen that the usual approach to minimize Kapitza resistance consists in increasing to surface area of the contact between liquid helium and copper. In dilution refrigerators where base temperature of a few mK are sought, a combination of continuous and step heat exchangers is used. In our

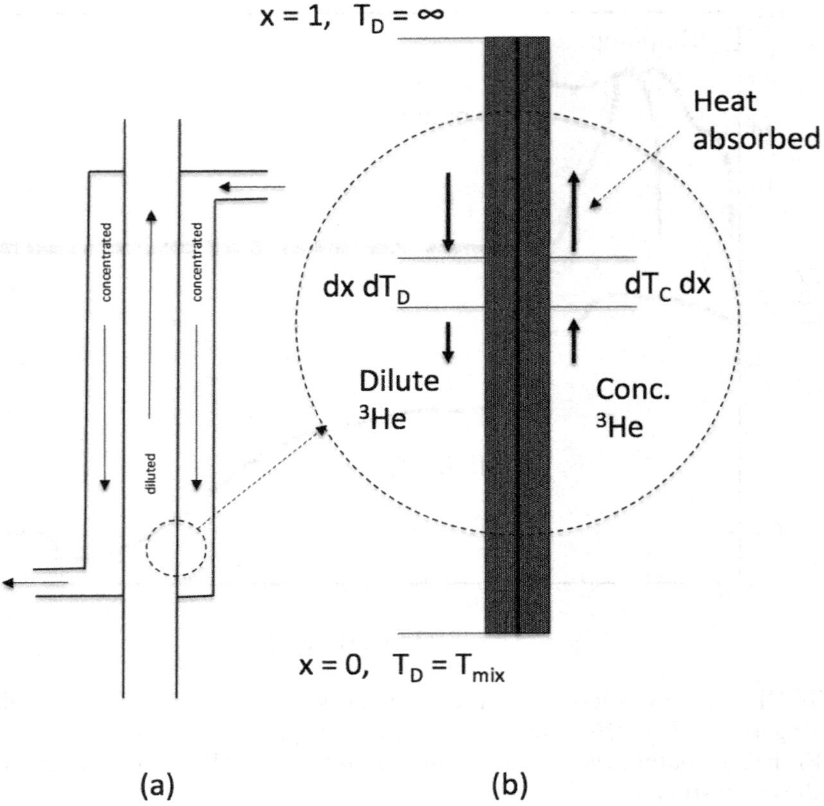

FIGURE 5.12: (a) Ideal continuous heat exchanger; (b) magnification of fluid boundary. Notice the direction of the thermal gradient is opposite to the direction of flow for both the dilute and the concentrated phases.

case, we will limit ourselves to the case of continuous heat exchangers only given that miniature dilution refrigerators, like Edel'man for example, are not meant to provide large cooling powers at very low temperatures. In addition, dilution systems using only continuous heat exchangers have demonstrated to be capable of reaching temperatures as low as 22 mK [93].

In a continuous heat exchanger used in our condensation pumped dilution refrigerator, the returning liquid ^3He to the mixing chamber is pre-cooled by the diluted ^3He (see Fig. 5.12). In an ideal heat exchanger we assume that both lateral heat conduction and viscous heating are negligible. No matter how complicated the heat exchanger in the dilution system is, it can be always represented by a single heat exchanger with area A with one side in contact with the dilute ^3He and on the other side with the concentrated ^3He [83].

Following [83] and [28] at temperatures $T < 50$ mK pure liquid ^{3}He and liquid ^{3}He dissolved in superfluid ^{4}He are Fermi liquids whose entropies can be expressed as:

$$S_{C,D}(T) = \gamma_{C,D} T \tag{5.20}$$

for the dilute ^{3}He we have already seen that $\gamma_D = 106\ JK^{-2}mol^{-1}$, and that for the concentrated we have $\gamma_C = 22\ JK^{-2}mol^{-1}$. We can now express the cooling power in the mixing chamber as:

$$\frac{\dot{Q}}{\dot{n}} = (\gamma_D - \gamma_C)T_m^2 \tag{5.21}$$

where T_m is the temperature of the mixing chamber and assuming that the returning ^{3}He has thermalized at T_m before entering the mixing chamber. This equation is useful because it sets an upper limit to the cooling power.

It is found empirically that the cooling power of the dilution refrigerator obeys a power law $\propto T_m^2$ only when the mixing chamber is at a temperature about 3 times higher than the base temperature, i.e., the temperature achieved with the minimal thermal input. The fundamental importance of the counter-flow heat exchanger can then be understood because it will pre-cool the returning ^{3}He to the mixing chamber. If the heat exchanger is not properly designed, the returning ^{3}He will bring additional avoidable thermal input reducing the cooling power. In this case, we need to modify Eq. 5.21 to consider the heat capacity of the returning ^{3}He, since it is at a higher temperature than T_m. Using the fact that the heat capacity of a Fermi liquid is equal to its entropy at low temperatures, we get

$$\frac{\dot{Q}}{\dot{n}} = (\gamma_D - \gamma_C/2)T_m^2 - \gamma_C \frac{T_C^2}{2} \tag{5.22}$$

Eq. 5.22 is not particularly useful since T_C depends on T_m and it is practically impossible to measure T_C in a running dilution system separately from the temperature of the diluted phase running up to the still. Takano [83] gives the proper equation:

$$\frac{\dot{Q}}{\dot{n}} = \frac{(\gamma_D - \gamma_C)[\gamma_D(f-1) - \gamma_C]}{\gamma_D(f-1) - \gamma_C(f+1)} \tag{5.23}$$

where f is a dimensionless factor equal to:

$$f = \exp\left(\frac{A/R_3}{\dot{n}\gamma_D\gamma_C}[(\gamma_D - \gamma_C)T_m^2 - 2\dot{Q}/\dot{n}]\right) \tag{5.24}$$

where R_3 is the coefficient of the T^{-3} dependence of the Kapitza resistance according to:

$$R_K = \frac{R_3 T^{-3}}{A} \tag{5.25}$$

Takano [83] gives more practical equations to avoid the exponential factor in Eq. 5.24:

$$\frac{\dot{Q}}{\dot{n}} = \left(\frac{\gamma_D}{2} - \frac{\gamma_C}{2}\right) T_m^2 - \frac{\gamma_D \gamma_C}{2} \frac{g \dot{n} R_3}{A} \tag{5.26}$$

where

$$g = \ln \left(\frac{(\gamma_D + \gamma_C) \left[(\gamma_D - \gamma_C) T_m^2 - \dot{Q}/\dot{n} \right]}{(\gamma_D - \gamma_C) \left(\gamma_D T_m^2 - \dot{Q}/\dot{n} \right)} \right) \tag{5.27}$$

Equations 5.26 and 5.27 can be solved numerically to properly design heat exchangers in dilution refrigerators. Takano [83] gives also the formulas for heat exchangers where the Kapitza resistance is proportional to $\propto T^{-2}$ where the heat exchangers are manufactured out of silver powder or evaporated silver. It has been shown that such heat exchangers have a Kapitza resistance between liquid and silver $\propto T^{-2}$.

5.2.2 Design considerations for a condensation pumped dilution refrigerator

The design of a dilution system starts usually with the requirements in terms of operating temperature T_m and thermal input to the mixing chamber $\dot{Q}$ together with practical assumptions on size, weight, dimension, etc. of the system. For example, in the design of a bolometric instrument devoted to continuum observational astrophysics at mm/sub-mm wavelengths, the mixing chamber usually cools the detectors and the optimal operating temperature, while the thermal input to the mixing chamber is dominated by electrical readout wires or support structures for the cold focal plane. Although there is no unique recipe for such a design, the necessary equations have been given in the previous sections and will be summarized here. Given T_m and $\dot{Q}$, Eq. 5.26 provides a relationship between the ^{3}He circulation rate, the γ coefficients of the diluted and concentrated phases, and the area A of the heat exchanger. In our case, the γ coefficients, respectively for diluted and concentrated phases, are $\gamma_D = 107$ JK^{-2}mol^{-1} and $\gamma_C = 23$ JK^{-2}mol^{-1}. There is a practical limit to the value of the circulation rate $\dot{n}$. In order to operate efficiently, the re-condensing temperature of the ^{3}He evaporating from the still should not exceed $\sim$ 400 mK. If the re re-condensing temperature is too high, then the returning ^{3}He carries too much heat thus decreasing the cooling power. In addition, the re-condensing temperature should always be much lower than the still temperature to have effective circulation. In the case of standard closed-cycle ^{3}He refrigerators like the ones described in Section 5.2, we see that the max temperature of 400 mK is achieved with a thermal input of $\dot{Q}_{cond} \sim 400$ μW. Using

$$\dot{n} = \frac{\dot{Q}_{cond}}{L_{3D}} \tag{5.28}$$

we obtain that the maximum circulation rate should be less than $\dot{n} < 10$ μmols^{-1}. Having established $\dot{n}$, we now need to size the heat exchanger. In the case of a continuous tube-in-tube heat exchanger, we need to estimate the Kapitza resistances between liquids and metals. For CuNi tubes, commonly used at 100 mK, the Kapitza resistance R_K has been measured to be constant at low temperatures and equal to $R_K = 0.01\ m^2K^4W^{-1}$ [52]. In a tube-in-tube heat exchanger there are two temperature discontinuities in series: liquid/metal and then metal/liquid. Therefore the total Kapitza resistance will be twice the value given above. The last parameter to be determined is the area A of the heat exchanger that depends on the diameters and lengths of the tubes used. For example, in a very compact configuration [84], we can use 60 mm length of CuNi tubes of diameters 2.3 and 3.3 mm with a total surface area of 43 cm^2. With these numbers we have a relationship between $\dot{n}$ and the expected thermal input. For example, if we expect 5 μW of thermal input, with a circulation rate $\dot{n} = 10$ μmol^{-1}, the minimum temperature achievable by our dilution system will be $T_m = 80$ mK. Notice that this temperature is the temperature measured **inside** the mixing chamber. To have a better estimate of the temperature at the outer surface of the mixing chamber, we need to calculate the ΔT introduced by the Kapitza resistance between the liquid in the mixing chamber and the metal. We see that even with a very modest surface area inside the mixing chamber of the order of 200 cm^2, using a value for the Kapitza resistance of $R_K = 0.02$ m^{2}K^{4}W^{-1} (interface between dilute ^{3}He and Copper) with a temperature of $\sim$ 100 mK we find that:

$$\Delta T = \frac{R_K \dot{Q}}{T_m A} = 5\ mK \tag{5.29}$$

Fig. 5.13 shows typical values of $R_K T^3$ for various interfaces as a function of temperature.

5.2.3 Other condensation pumped dilution refrigerator designs

We will now describe a few designs of successfully built condensation pumped dilution systems. Although all the designs are based on the same basic principles, many have some interesting solutions that exploit the portability and simplicity. As we mentioned earlier in the book, the Edel'man [24] system is the first prototype that has inspired a series of designs.

5.2.3.1 Quasi-single shot design

Edel'man [25] has proposed a design of a condensation pumped dilution refrigerator that has two advantages: (a) no continuous heat exchanger and (b) no ^{3}He return to the mixing chamber. An important difference consists in

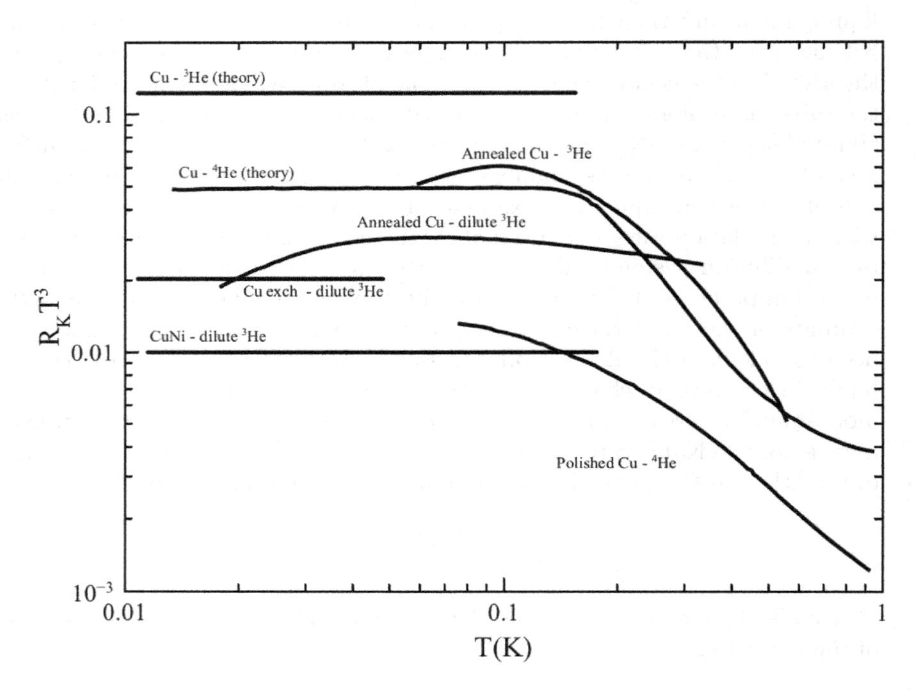

FIGURE 5.13: Kapitza resistance for various solids with liquid ^{3}He and ^{4}He (data from [52])

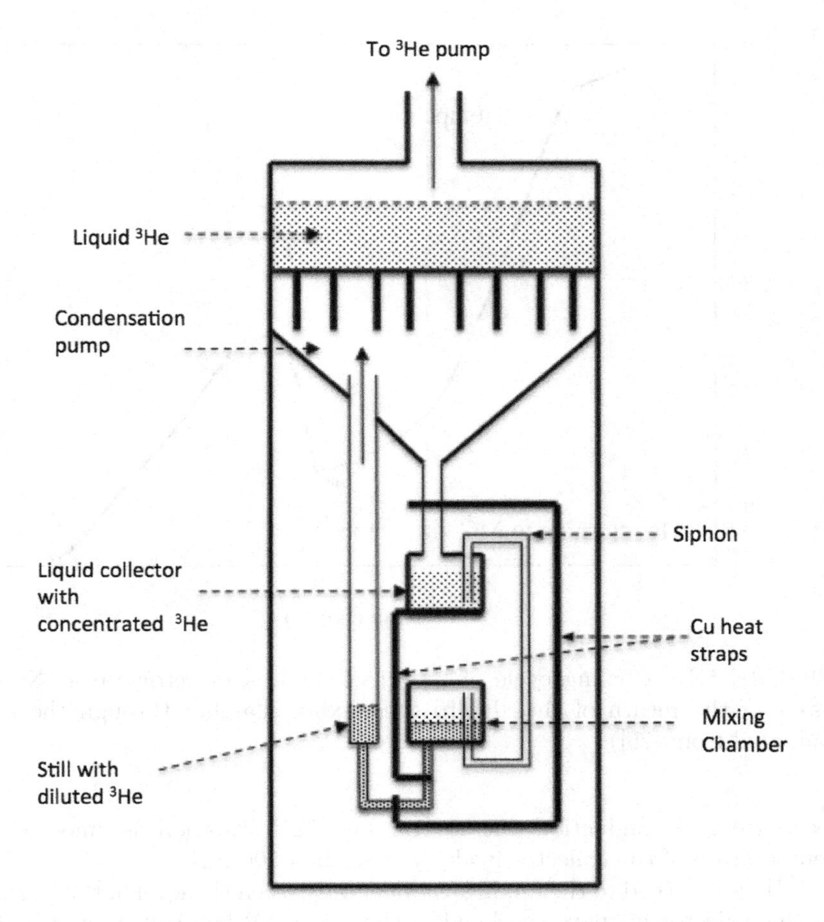

FIGURE 5.14: Quasi-single shot design (from [25])

having the still positioned at about the same level as the mixing chamber. This ensures that the refrigerator uses the minimum amount of liquid ^{4}He which should result in a shorter pre-cooling time from 400 to 500 mK to the lowest operating temperature.

As in the classic Edel'man design, in the system shown in Fig. 5.14, the ^{3}He evaporated from the still is condensed at 400 to 500 mK and drained down into a collector chamber. In this way, the liquid ^{3}He is not returned immediately into the mixing chamber thus removing an important source of heat. A small CuNi tube shaped as a siphon connects the collector chamber to the mixing chamber. The outer diameter of this tube is 0.7 mm with 0.2 mm thick wall. The inner diameter of 0.3 mm has been chosen to be smaller than the capillary length for ^{3}He. This solution guarantees that in the bottom part of the siphon there is always a plug of liquid ^{3}He which prevents an

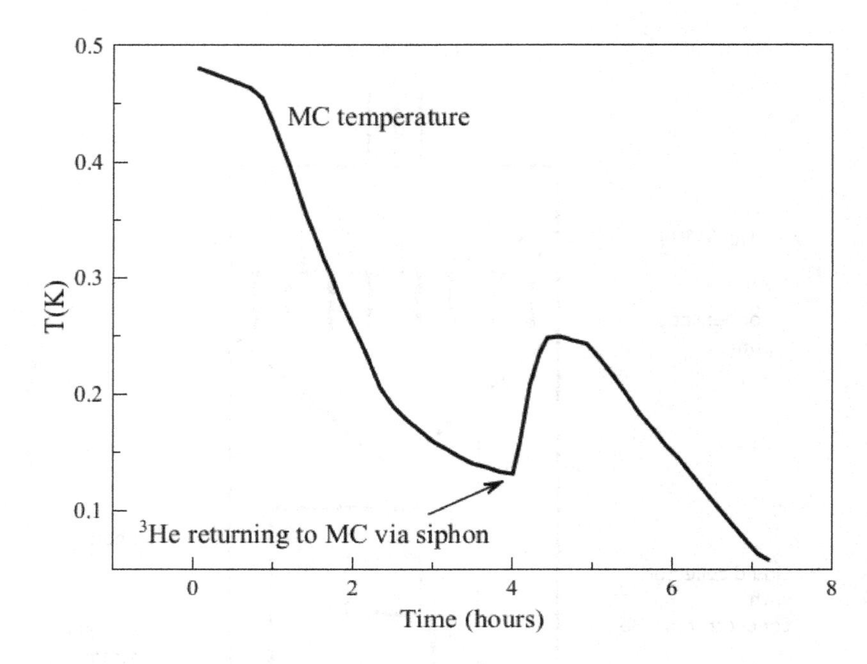

FIGURE 5.15: Cooling cycle of the quasi-single shot refrigerator. Note the onset of the return of the ^{3}He to the mixing chamber through the siphon (adapted from [25]).

extra thermal conduction due to ^{3}He gas. This situation assumes that the temperature of the collector is always less than 500 mK.

^{3}He is returned to the mixing chamber only when the level of the liquid ^{3}He in the collector is above the bend of the siphon. When this happens, all the liquid content in the collector is returned in the mixing chamber with a sudden thermal input. If properly designed, this burst of liquid ^{3}He will happen after several hours of operations. Having removed, for most of the time, the heat of the returning ^{3}He, this refrigerator runs as a normal dilution system when the return ^{3}He is shut off. It is well known that, in such conditions, the refrigerator runs at the lowest temperature.

Note that the collector is thermalized through a heat strap to a convenient point in the tube connecting the mixing chamber to the still. The returning ^{3}He from the condensation pump is also thermalized to another convenient point in the same tube connecting the mixing chamber to the still. In Fig. 5.15, a cycle of the fridge is shown. It is interesting to see the time at which the siphon is returning ^{3}He to the mixing chamber with the relatively small jump in mixing chamber temperature ($\sim$ 150 mK). It is important to notice that this refrigerator will run as long as the ^{3}He pot cooling the condensation

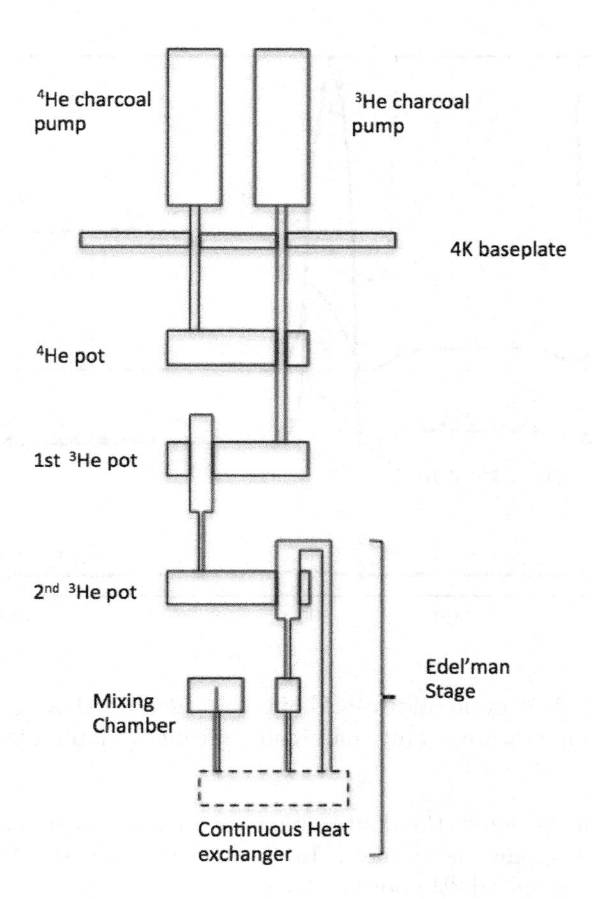

FIGURE 5.16: Continuous operating condensation pumped dilution system (adapted from [36])

pump is in operation. When ^{3}He runs out in this pot, the whole system needs to be re-cycled.

5.2.3.2 *A continuously operated dilution system*

Herrmann in 2005 has proposed an interesting evolution of the original system [36]. As we have noticed in the previous section, all the condensation pumped dilution refrigerators will run as long as the condensation pump is kept at its lowest temperature (typically $\sim$ 400 mK). In the system shown in Fig. 5.16, we see a standard Edel'man system — with the modification of the still at about the same level as the mixing chamber. As we already know, the Edel'man fridge runs as long as 300 mK is present at the condensation pump. If the dilution stage was attached directly to the first ^{3}He stage, when the ^{3}He

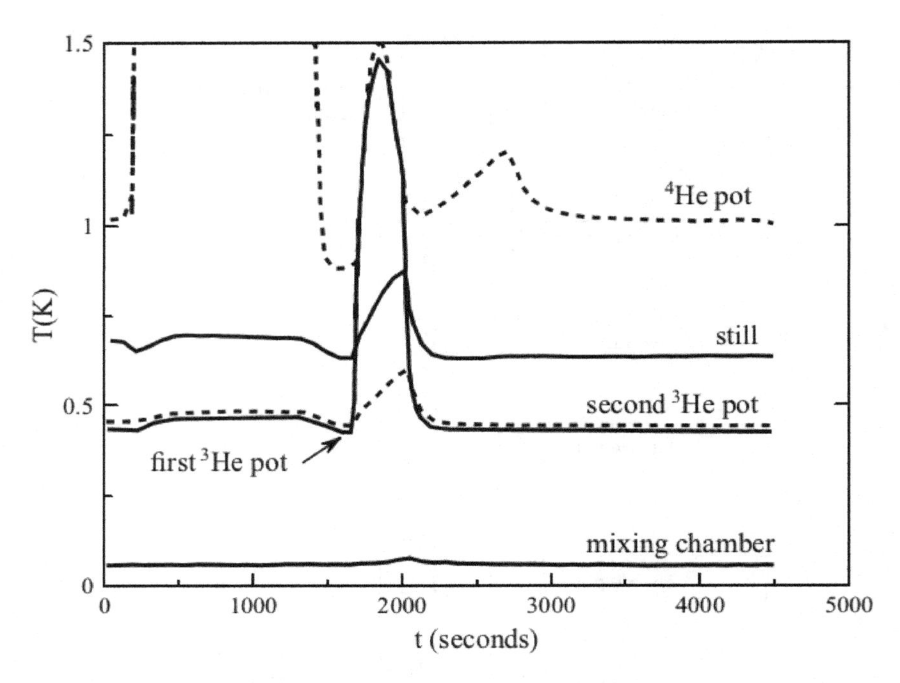

FIGURE 5.17: Recycle cold-to-cold of the refrigerator showing that the mixing chamber temperature is only marginally affected (adapted from [36])

pot runs out of liquid, the double stage sorption cooler needs to be re-cycled effectively stopping the dilution. In the system schematically drawn in Fig. 5.16, a second sealed ^{3}He pot is added.

This pot is connected to the first pot through a tube and a condenser. During the normal operation of the dilution stage, the heat of condensation of the ^{3}He coming from the still evaporates some liquid from the 2nd pot. The gas coming from the 2nd pot is re-condensed by the 1st pot. When liquid ^{3}He runs out in the first pot, there is enough liquid ^{3}He in the second pot to keep the dilution running while the double-stage fridge is re-cycled due to the relatively large specific heat of liquid ^{3}He. If the amount of ^{3}He in the second circuit is sized properly, the double-stage fridge can be recycled without effectively stopping the dilution. One of the most attractive features of this system consists in the little amount of He gas in the system: the two charcoal pumps contain 0.15 and 0.05 mol of respectively ^{4}He and ^{3}He. The second sealed ^{3}He circuit contains 0.1 mol. Finally, 0.04 mol of a 25-30% solution of ^{3}He in ^{4}He is in the mixture. The fact that relatively little gas is used in this system allows for short re-cycling time of the double stage refrigerator ($\sim$ 10 minutes).

Fig. 5.17 shows a re-cycling of the fridge. The large deviation in temperature of the various components is evident during regeneration while the mixing chamber stays relatively stable. An increase in temperature in the mixing chamber is evident only when the still is approaching $T = 0.8$ K mainly because ^{4}He start circulating thus effectively reducing the circulation of ^{3}He.

5.2.3.3 Dilution refrigerators with charcoal pumps - I

In 1984, Mikheev et al. [63] have proposed the use of charcoal pumps to keep the ^{3}He always cold and to pump efficiently the ^{3}He gas from the still. They built a system using two charcoal pumps working alternatively in desorption/absorption to make the dilution process continuous.

With reference to Fig. 5.18, while one of the two charcoal pumps pumps the ^{3}He over the still, the other is heated to desorb the ^{3}He collected during its previous pumping cycle. The desorbed ^{3}He is liquefied in the 1 K pot (kept at 1 K by pumping over the liquid ^{4}He bath). The liquid ^{3}He is collected in the collector by the action of gravity. The authors claim that the circulation speed is controlled by the saturated vapour pressure in the collector. The liquid ^{3}He from the collector is then passed through the still for additional precooling before entering the heat exchanger and the mixing chamber through a throttle. The ^{3}He pressure before the throttle is the sum of the ^{3}He hydrostatic pressure plus the ^{3}He vapour pressure inside the collector. If the collector is around 1.2 K, then the hydrostatic pressure is negligible with respect to the vapour pressure inside the collector. In this way, even though the ^{3}He level is changing in the collector due to the alternate regeneration of the charcoal pumps, the vapour pressure still dominates the circulation rate. If the collector temperature is constant, then the circulation rate is constant and the mixing chamber temperature will be constant during the switch between charcoal pumps. The switching between the two ^{3}He charcoal pumps is achieved by operating on the cold valves 1, 2, 3 and 4. For example, if charcoal pump 1 is in desorbing phase and charcoal pump 2 is in absorbing phase, then valves 1 and 4 are closed while valves 2 and 3 are open. Reversing the charcoal pump actions is achieved by reversing the status of all the valves.

The system described has certainly some attractive features like, for example, the use of the strong pumping action of the charcoal pumps and the completely cold circulation of ^{3}He. However, it must be noted that the operations depend on the action of mechanical valves in the cold stages. In addition, the 1 K pot still needs an external mechanical pump to keep the temperature needed to re-condense the ^{3}He.

5.2.3.4 Dilution refrigerators with charcoal pumps - II

An improved design of a continuous condensation pumped dilution refrigerator is described in Sivokon et al. [76]. A schematic drawing of the principles of operation of this fridge is shown in Fig. 5.19.

The idea behind this refrigerator is simple: there are two different condensantion pumps connected to two independent ^{3}He pots. The ^{3}He gas evaporat-

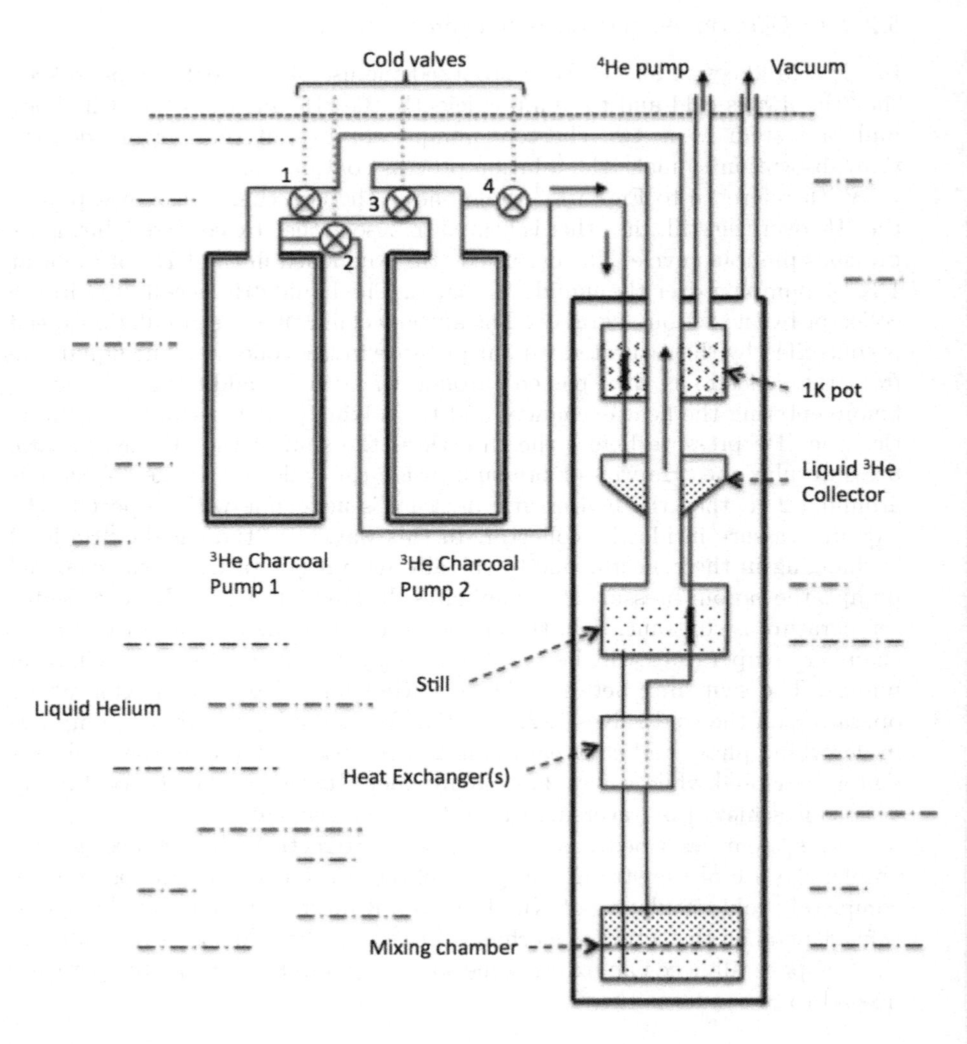

FIGURE 5.18: Charcoal pump operated condensation pumped dilution refrigerator (adapted from [63])

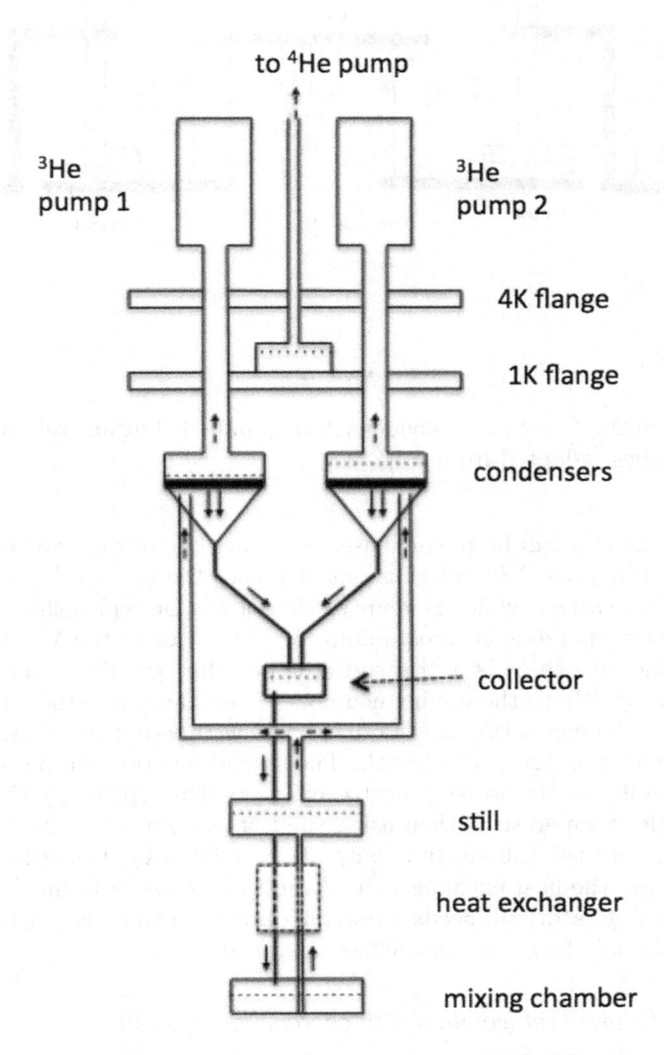

FIGURE 5.19: Continuous condensation pumped dilution refrigerator (adapted from [76])

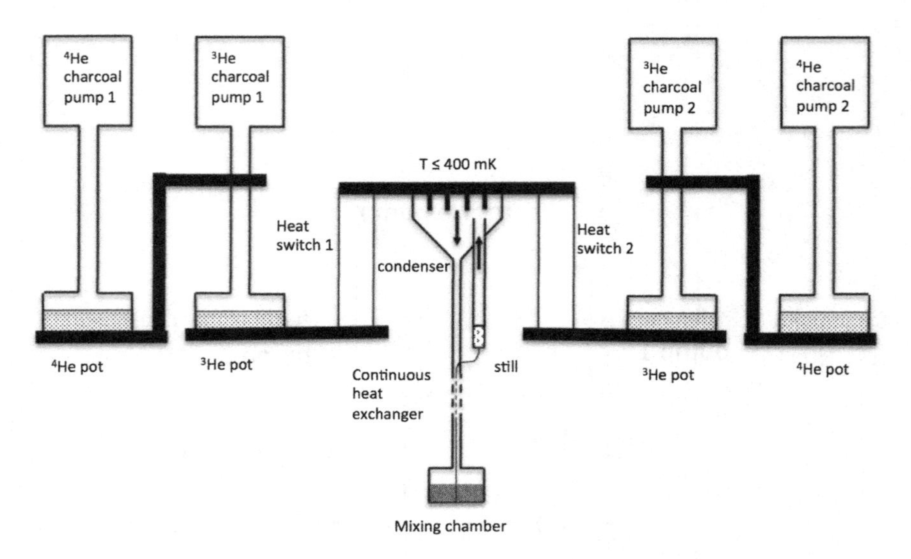

FIGURE 5.20: Continuous condensation pumped dilution refrigerator with heat switches (adapted from [84])

ing from the still can be re-condensed on either one of the two cold surfaces. However, when one ^{3}He pot runs out of liquid, the gas can be re-condensed on the other surface while the empty ^{3}He pot can be replenished by regenerating its corresponding charcoal pump. As in the case of the Mikheev system, in this case also there is a ^{3}He collector pot that stabilizes the pressure of the returning ^{3}He to the mixing chamber. They have also studied two different heat exchanger sections. The first has one tube-in-tube continuous heat exchanger that is 120 cm in length. The second has two continuous heat exchangers followed by two step heat exchangers. Unsurprisingly they report a much colder temperature when using the last configuration ($T_{mix} \sim 12$ mK) compared with the tube-in-tube only ($T_{mix} \sim 50$ mK). The still, the mixing chamber and the heat exchangers are practically standard units.

This refrigerator still needs a pumping line from the 1 K pot to the pump keeping the 1 K flange at constant temperature.

5.2.3.5 Dilution refrigerators with charcoal pumps - III

So far, all the previous systems rely on some external connection to the cryostat, for pumping, compressing ^{3}He or both. A completely sealed system able to continuously keep the mixing chamber at the lowest temperature without any external connection has been discussed by Teleberg [84]. A schematic view of this refrigerator is shown in Fig. 5.20.

There are two main additions with respect to previous systems of this kind: the switching between the two ^{3}He pots is achieved by means of gas

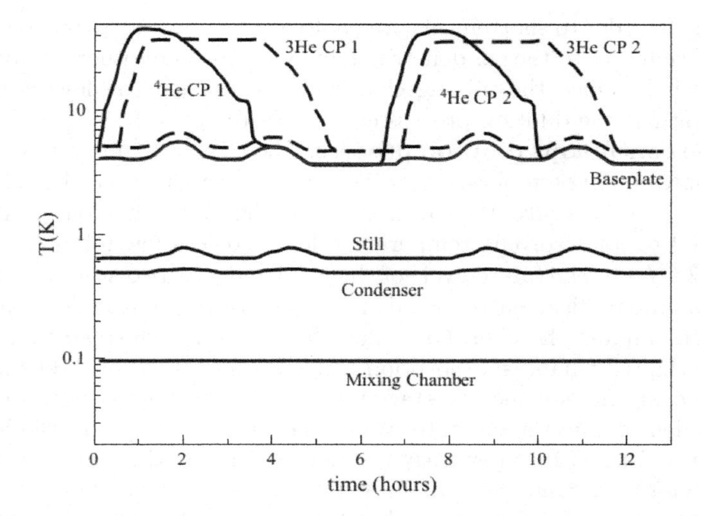

FIGURE 5.21: Cooling cycle of Teleberg's refrigerator (adapted from [84])

heat switches and there are no external connections since all the needed temperatures are generated by sealed sorption coolers.

In Fig. 5.21 it is possible to follow the actions of the charcoal pumps and heat switches (not indicated) to maintain the mixing chamber around 100 mK. The cycle of this system starts when all the components are at the base temperature of ~ 4 K — either from a liquid helium bath of the second stage of a Gifford-McMahon or Pulse Tube Cooler.

Mixing Chamber pre-cooling phase. Both ^{4}He charcoal pumps 1 and 2 are heated to approximately 50 K. The ^{4}He gas in the sealed circuits will be desorbed by the charcoal pumps and will start to liquefy when the inner pressure exceeds the saturated vapour pressure at ~ 4 K. After a brief delay of a few minutes, the ^{3}He charcoal pumps 1 and 2 are also heated to their desorbing temperature (~ 40 to 50 K). We have seen in Section 4.1 that even if the base temperature is higher, we can expect that a fraction of the gas will liquefy. Once the liquefaction of ^{4}He is completed — usually after keeping the charcoal pumps at the desorbing temperature of ~ 40 to 50 K — the charcoal pump heat switches can be closed effectively connecting the charcoal pumps to the 4 K flange. The charcoal pumps will start pumping over the liquid ^{4}He to reach quickly a base temperature of ~ 1 K. The two switches connecting the ^{3}He pots can be closed in order to pre-cool the re-condensing pot in the dilution circuit. At this point, we can close the heat switches (not indicated in Fig. 5.20) connecting the ^{3}He charcoal pumps to the base plate. After a few minutes both ^{3}He pots will reach their base temperature of ~ 300 mK.

The mixing chamber, being very well isolated, would need a very long time to reach the required $T_{mix} < 400$ mK needed to efficiently start the dilution process. In order to shorten this pre-cooling time, it is necessary to connect the mixing chamber to the condenser flange by means of another passive or active heat switch. Once the mixing chamber has reached the lowest pre-cooling temperature, the dilution process can be started.

The continuous dilution process. The mixing chamber has reached the lowest pre-cooling temperature, while the two ^{3}He pot reached the lowest temperature of $\sim$ 300 mK. We can now follow Fig. 5.21: charcoal pump 1 (^4He) is heated to its desorbing temperature followed in a few minutes by charcoal pump 2 (^3He). The full re-cycle of the left-hand side two-stage sorption cooler has a minimal effect on the condenser temperature and little or no effect at all on the mixing chamber. Once the left-hand side two-stage fridge has been cycled, the right-hand side sorption fridge can now be cycled. Obviously, during the cold phase of the two-stage refrigerator, the heat switch connecting to the condenser is in the on state. While the fridge is re-cycled, the heat switch is in the off state. It is now easy to understand why these heat switches need to have high on conduction and low off conduction. The high on conduction ensures a minimal ΔT between the ^{3}He pots and the condenser, while a high off conduction ensures a small heat leak to the condenser when the ^{3}He pot is brought to $\sim$ 1 K during the recycle phase.

5.3 SINGLE-SHOT 100 MK SYSTEMS

We already mentioned that if in a conventional dilution refrigerator we close the return ^{3}He line, the mixing chamber goes to its minimal temperature. This is simply explained by considering that no liquid ^{3}He is returning to the mixing chamber thus eliminating the additional thermal input. Very few research teams have tried to design and build a dilution system where the ^{3}He is not re-circulated back to the mixing chamber (single-shot dilution system, SSDS). One major disadvantage of such a system consists in the limited running time. Once the ^{3}He in the mixing chamber is exhausted, the dilution is stopped and the system needs to be re-cycled. One obvious advantage, on the other hand, consists in the absence of a heat exchanger making the SSDS less complex. Such a system can be used where, for example, continuous cooling is not needed. Solid state experiments where there is the need to change samples regularly, for example, will benefit from such a system providing that the cooling power and the length of the cold state is compatible with the specifications. Another area of use could be astrophysics where, in certain circumstances, the simplicity of the SSDS is enough to justify the loss of observing time. Dilution systems attached to telescopes, with stringent requirements in terms of complexity, are not compatible with pumping and compressed gas lines, gas handling systems, etc. that characterize **standard** dilution system. Two-axes tilting, typical of certain telescope observations, can also be a serious requirement that can be difficult, if not impossible, to satisfy.

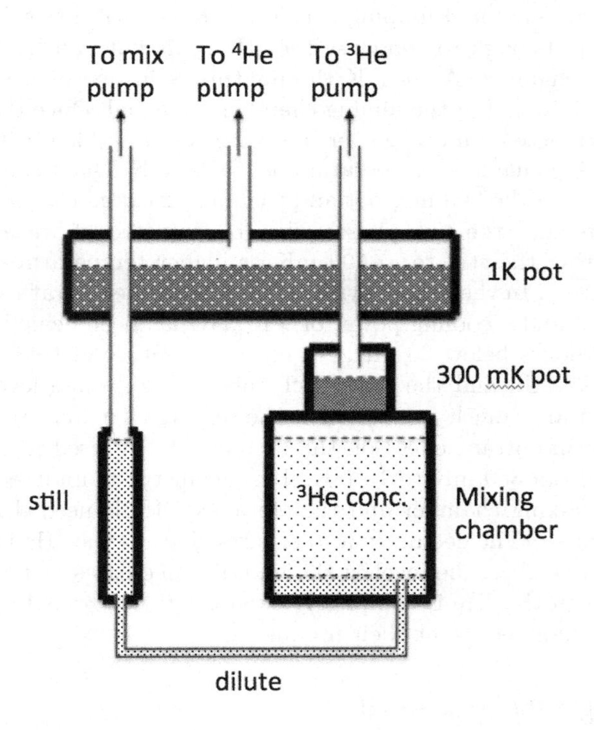

FIGURE 5.22: Schematic of the Roach et al. single-shot dilution system [69]

5.3.1 Single-shot system - I

A SSDS that makes use of 1 K pot is described in [69] and a schematic view is shown in Fig. 5.22. This system is assumed to be in a sealed vacuum chamber surrounded by a liquid helium bath. Exchange gas can be admitted to pre-cool and thermalize everything at 4.2 K. ^{4}He gas can be admitted in the 1 K pot. If the pressure of the gas is higher than the saturated vapour pressure (in this case $\sim$ 1 bar), the gas liquefies and drops by gravity at the bottom of the 1 K pot. Once enough ^{4}He has been collected, ^{3}He can be admitted in the sealed circuit and the mixture ^{3}He + ^{4}He gas can be admitted in the sealed mixture circuit. Pumping over the 1 K pot will lower its temperature to $\sim$ 1 K, low enough to condense the ^{3}He in its pot and the mixture in the still/mixing chamber. At $\sim$ 1 K the mixture is in one phase and the liquid levels in the still and in the mixing chamber are equal. Once the condensation of ^{3}He is completed, pumping over the 300 mK pot will lower its temperature, and the mixing chamber temperature, to $\sim$ 300 mK. Due to the good thermal contact between the 300 mK pot and the mixing chamber, the mixture liquid will separate into the two phases: the concentrated above and the diluted below. Heating the still to $\sim$ 600 mK or higher temperature, will start the dilution system. In the Roach system, a minimum temperature of 15 mK has been reached and a cooling power of >10 μW has been achieved. The system runs for 10 hours below 20 mK. An essential feature of this system consists in having the still and the mixing chamber at the same level and the still cross-section area much smaller than the mixing chamber cross-section area. This has an important effect on the amount of ^{3}He used in the mixture for pre-cooling from 300 mK to 15 mK. This geometry minimizes the amount of ^{4}He in the mixing chamber and therefore less ^{3}He is needed to pre-cool the mixing chamber. The geometry also ensures that, as the ^{3}He is depleted from the mixing chamber, the level in the still also decreases at the same rate. In this way, when the ^{3}He is completely exhausted, the levels in the still and in the mixing chamber are at their minimum.

5.3.2 Single-shot system - II

Another SSDS has been described in [60] and is shown schematically in Fig. 5.23.

This SSDS has been designed to be operated at tilt angles as high as 60° from the vertical and capable of working also with a range of rotation angles. Since this SSDS is designed to operate in a dry system, the pre-cooling of the cold pots is achieved with a series of convection assemblies. Once the base plate has reached the lowest temperature, which in a dry system can be as low as 2.5 K, all the cryopumps can be kept at desorbing temperatures ($\sim$ 50 K). The ^{3}He circuit, for example, has two independent loops for the gas. The first loop, between the base plate and the 1 K pot, provides pre-cooling of the 1 K pot and condensation of the inner ^{4}He in the pot. The second

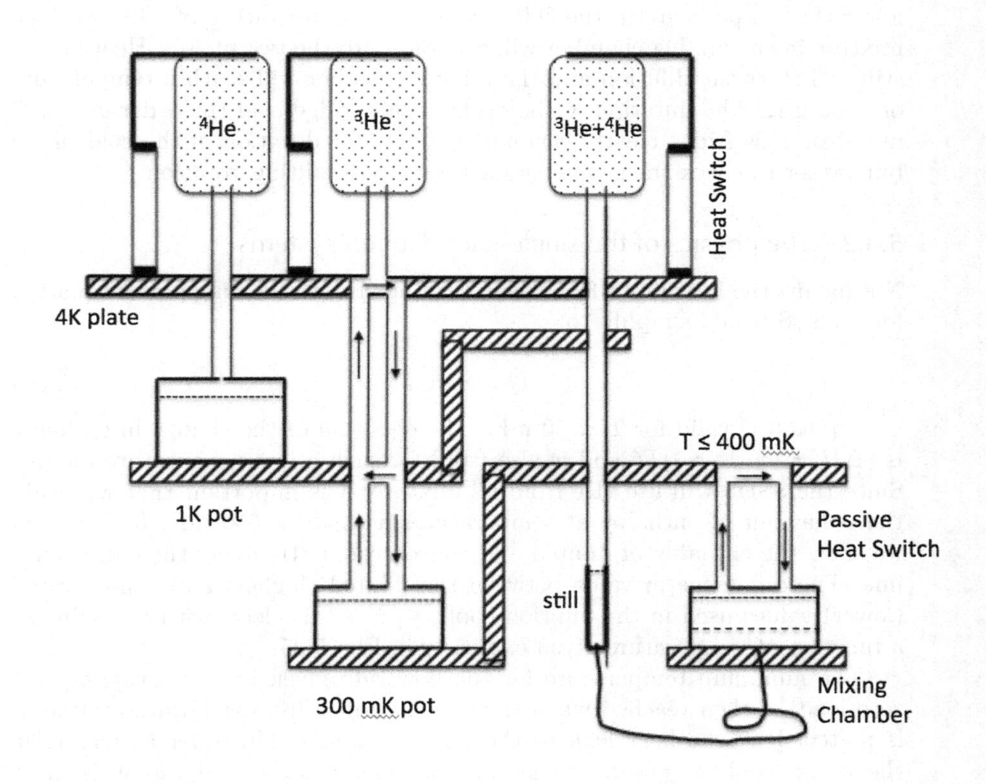

FIGURE 5.23: Schematic of the Melhuish et al. single-shot dilution system [60]

loop provides pre-cooling for the ^{3}He pot. Once all the ^{4}He in the 1 K pot is condensed, the ^{4}He cryopump can be cooled, through its heat switch, and the 1 K pot will reach quickly $\sim$ 1 K. At this temperature, practically all the ^{3}He will be condensed in the ^{3}He pot. The mixture will also condense, in a single phase, in the mixing chamber and in the still. As in the Roach system, also the Melhuish system has the still and the mixing chamber at the same level. Once all the ^{3}He is condensed, its corresponding cryopump can be cooled through activation of its heat switch. The cooling of the cryopump will lower the temperature in the 300 mK pot to a temperature of 400 mK. The mixture in the mixing chamber will separate into the two phases. Heating the still will start the dilution and the mixing chamber will reach a temperature of $\sim$ 65 mK. The duration of the cycle of this fridge is of the order of $\sim$ 100 minutes. This fridge clearly was not designed for duration of the cold phase but rather to check the geometry for the extreme tilting/rotations.

5.3.3 The physics of the single-shot dilution systems

Not having the heat input from the returning ^{3}He, the cooling power equation for a SSDS would simplify to

$$\dot{Q} = 84\dot{n}_3 T^2 \tag{5.30}$$

Eq. 5.30 is valid for $T < 50$ mK. The equation of the change in enthalpy is $(\Delta H = (\gamma_D - \gamma_C)T^2$ and is also valid for such a low temperature regime. Since the SSDS will use ^{3}He from its mixture, it is important that we study the behaviour of enthalpy at temperatures as high at 500 mK. In Fig. 5.24 we show the enthalpy of diluted and concentrated ^{3}He along the coexistence line. The difference in value between the diluted (higher) and concentrated (lower) values used in the dilution cooling process is clear, with the value as a function of temperature given explicitly in Fig. 5.25.

The minimum temperature for the flow rate given by the cryopump (at zero load) is then reached when the cooling power from the dilution process is matched by the heat leak to the mixing chamber. In order to maximize the useful cooling power, the mixing chamber (and indeed the still) must be extremely well thermally isolated. One important design issue consists in sizing the system to use as little ^{3}He as possible in the mixture to pre-cool the mixing chamber. In this way, the run time can be maximized, having left in the mixing chamber the maximum amount of ^{3}He.

Suppose we want to calculate the amount of ^{3}He to be removed to cool the mixing chamber from T_0 to T_1. We would expect, and we would be right, that the lower the starting temperature T_0, the less ^{3}He will be used. We have that [84]:

$$-dn_3 = \frac{dQ}{H_{3D} - H_3} \tag{5.31}$$

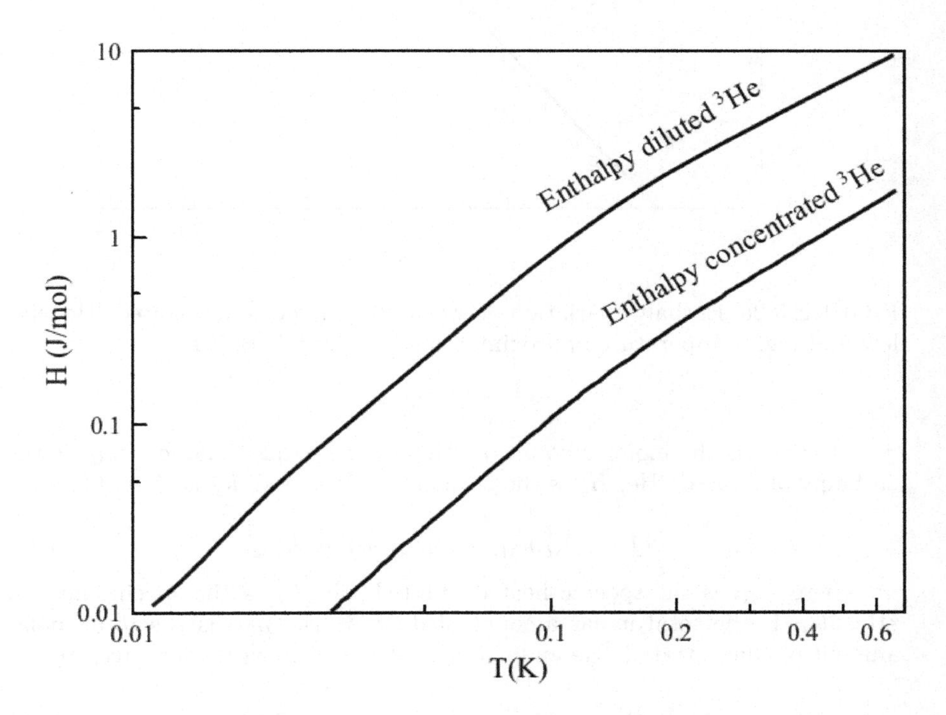

FIGURE 5.24: Enthalpy of diluted and concentrated ^{3}He along the coexistence line [52]

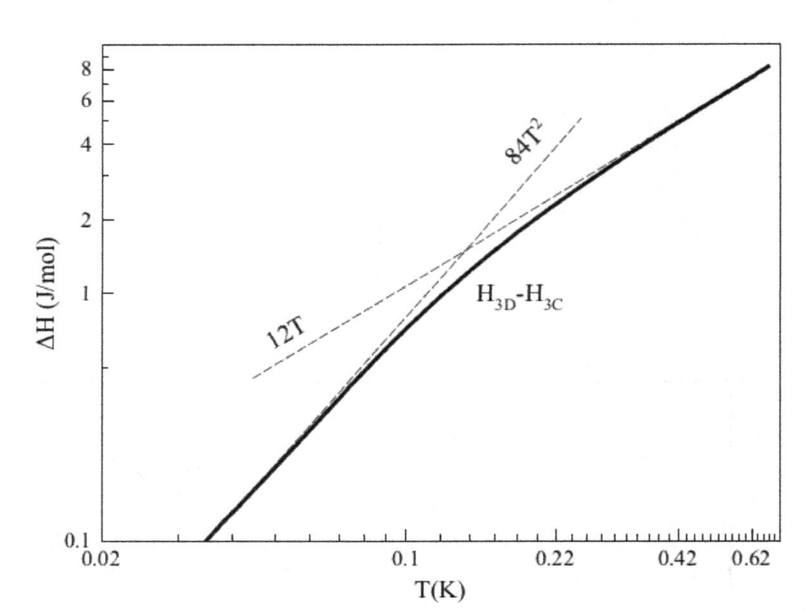

FIGURE 5.25: Enthalpy variation between diluted and concentrated ^{3}He with low and high temperature approximations (adapted from [84])

where n_3 is the molar amount of ^{3}He in the mixing chamber, H_{3D} is the enthalpy of diluted ^{3}He, H_3 is the enthalpy of ^{3}He, and dQ is given by:

$$dQ = (C_{3D}n_{3D} + C_3 n_{3C})dT + dQ_{sep} \tag{5.32}$$

where C_{3D} is the specific heat of diluted ^{3}He, C_3 is the specific heat of ^{3}He, n_{3D} is the total molar amount of diluted ^{3}He, n_{3C} is the total molar amount of concentrated ^{3}He and dQ_{sep} is the heat of separation given by:

$$dQ_{sep} = (H_{3D} - H_3)\frac{d}{dT}(x_{3D}n_D)dT \tag{5.33}$$

where x_{3D} is the ^{3}He concentration in the diluted phase and n_D is the total number of moles in the diluted phase. Enthalpy and entropy balances for ^{3}He + ^{4}He mixtures can be obtained through numerical integration of either experimental specific heat capacity data (Radebaugh [67] and Kuerten [45]) or by using polynomial fit (Chaudhry [12]).

The pre-cooling ^{3}He fridge determines the minimum starting temperature T_0. The molar flow rate of ^{3}He can be determined by the pumping speed of the charcoal pump (in the mixture circuit) and the temperature of the still. We can then calculate the rate of crossing of ^{3}He at the phase separation with the associated cooling. To this cooling we have to add the structural heat leaks through the supports, the tubing, etc. plus any heat dissipated on the

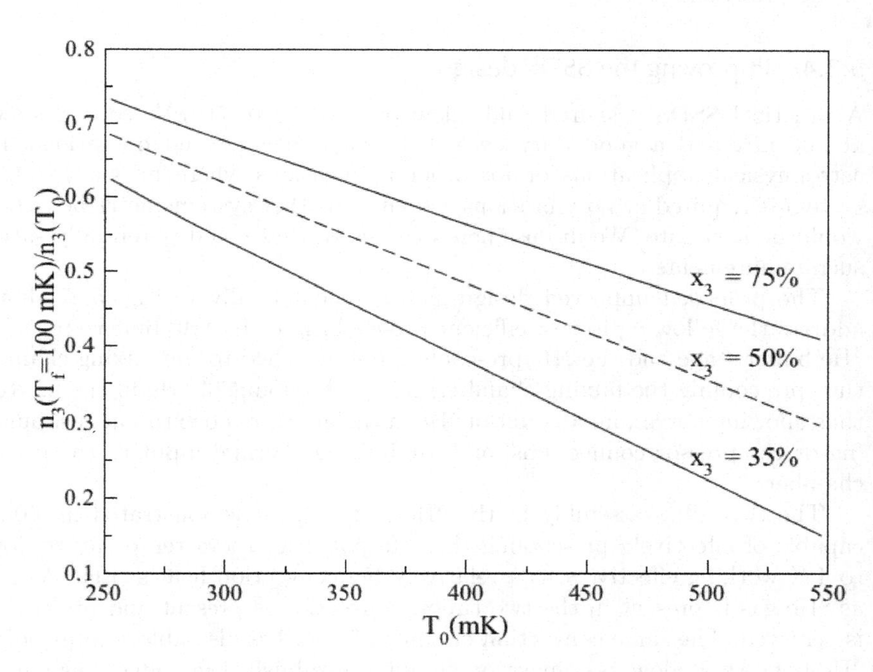

FIGURE 5.26: Fraction of ^{3}He left when pre-cooling from T_0 to 100 mK assuming initial fractions of 35%, 50% and 75% [60]

load connected to the mixing chamber. The calculation proceeds numerically for each time step, with molar quantities in each phase tracking the helium transfer between phase boundary. The mixture temperature is then adjusted according to ΔQ and the heat capacity of the mixture from the data or the polyfit.

In Fig. 5.26 it is possible to determine how much ^{3}He will be left after the pre-cooling as a function of the starting temperature of the dilution (T_0) for 3 different initial concentrations of ^{3}He in the mixture. It is clear that a large initial fraction of ^{3}He will give the maximum fraction of ^{3}He left at the end of the pre-cooling.

5.3.4 Improving the SSDS design

A practical SSDS design should allow to have 10 to 50 μW cooling power at 100 mK and a good duty cycle between re-cycling and operations. For astrophysical applications or for other applications where no gas handling system is required or no vibrations are vital, neither systems described above would be adequate. We discuss here one design that could potentially satisfy such requirements.

The proposed improved design shown schematically in Fig. 5.27 should address the following issues: efficient pre-cooling of the ^{4}He buffer stage, the ^{3}He buffer stage and the ^{3}He pre-cooling pot attached to the mixing chamber thus pre-cooling the mixing chamber; minimal amount of ^{4}He in the mixture thus allowing maximum amount of ^{3}He in the mixture; no external gas, pumping or compressor connections; and minimizing thermal input to the mixing chamber.

The two-tubes assembly in the ^{3}He fridge 1, as demonstrated in [60], is capable of effectively pre-cooling the ^{4}He pot from room temperature down to 4 K working effectively as a relatively big convection heat switch. As long as ^{3}He gas is present in the two tubes, convection is present and pre-cooling is achieved. The same convection circuit in fridge 1 is also able to pre-cool its ^{3}He pot. An analogous convection circuit is established in fridge 2 as long as ^{3}He gas is present. Therefore, in order to reach 4 K in all the subsystems, all charcoal pumps should be kept above 50 K during initial cooldown from room temperature.

The mixing chamber, with its mixture gas inside, is pre-cooled by thermal conduction, through the common copper, with the second ^{3}He circuit. Once the temperature of all cold parts is around 4 K, we can turn off the charcoal pumps in sequence from left to right. The ^{4}He pot is now containing liquid ^{4}He and therefore cooling its charcoal pump will rapidly bring the temperature of the ^{4}He pot to around 1 K. Since the first ^{3}He charcoal is still hot, the ^{3}He gas in the circuit will condense and collect in the first ^{3}He pot at a temperature slightly higher than $\sim$ 1 K. The second ^{3}He circuit is also pre-cooled via convection through the two-tubes assembly. As in the first circuit, also this second circuit will condense liquid ^{3}He in the second ^{3}He pot. The

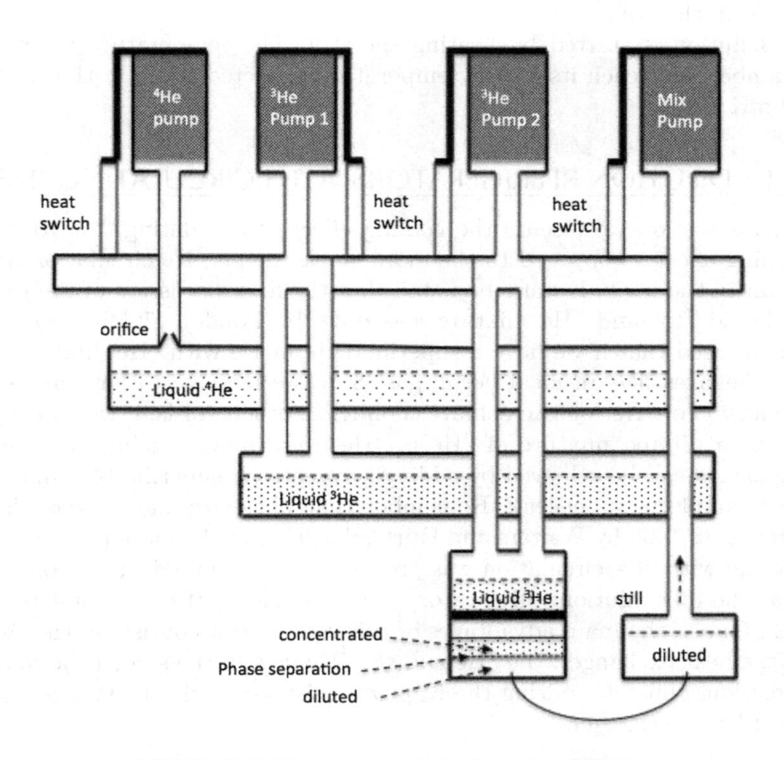

FIGURE 5.27: Improved design of a SSDS

mixing chamber will also pre-cool around $\sim$ 1 K through conduction across the common copper. The mixing chamber and the still will then fill with liquid mixture which, being above phase separation temperature, will still be in one single phase.

Once all the subsystems are around 1 K, we can now cool the two ^{3}He charcoal pumps bringing their respective pots around 300 mK. When the mixing chamber reaches its final pre-cooling temperature, expected to be of the order of $\sim$ 300 mK, phase separation has already been achieved and the last cryopump (the mix pump) can now be cooled starting pumping over the liquid in the still.

Dilution is started by heating the still. The temperature in the mixing chamber will reach its lowest temperature expected to be in the range 20 to 50 mK.

5.4 DILUTION REFRIGERATORS WITH CIRCULATING ^{4}HE

In this section we discuss the cooling effect of circulating ^{4}He to achieve a cooling effect as opposed to the more *conventional* ^{3}He circulation discussed so far in the book. It must be noted that the first discussion of the properties of liquid ^{3}He and ^{4}He mixture was done by London [50] in 1951. London has noticed that if we have a superfluid He phase with ^{3}He diluted in it, the ^{3}He behaves like an ideal Fermi gas. So if we manage to have an *isentropic* expansion of ^{3}He, we can achieve cooling. In other words, if we add superfluid He to a diluted mixture of ^{3}He in ^{4}He, we achieve cooling. The isentropic expansion can be achieved by adding zero-entropy superfluid helium through, for example, a super-leak. Remember that phase separation was discovered later on in 1956 by Walters and Fairbanks [92] and the more powerful dilution cooling with ^{3}He circulation was proposed and exploited. It was only in 1971 that the first dilution refrigerator using circulating ^{4}He was built by Taconis [82]. One of the main advantages of this refrigerator consists in the simplicity of the heat exchanger: the ^{4}He and the ^{3}He flow in the same tube in opposite directions thus minimizing the Kapitza resistance and achieving a practically ideal heat exchanger.

5.4.1 Superfluid ^{4}He pump

Superfluid helium, i.e., liquid helium at a temperature $T < T_\lambda$, is a Bose Einstein Condensate and therefore is described by a macroscopic wave function. The liquid therefore shows no viscosity unless it flows above a critical velocity. In Fig. 5.30 we see the viscosity of liquid helium as a function of temperature. Above the transition temperature $T_\lambda = 2.17$ K the viscosity shows an increase in viscosity with decreasing temperature reaching a maximum while approaching T_λ. Below T_λ the viscosity precipitates to zero. The viscosity of liquid helium is an interesting and complex issue and we refer the reader to the specialized literature [86, 95]. In addition to viscosity, knowing entropy

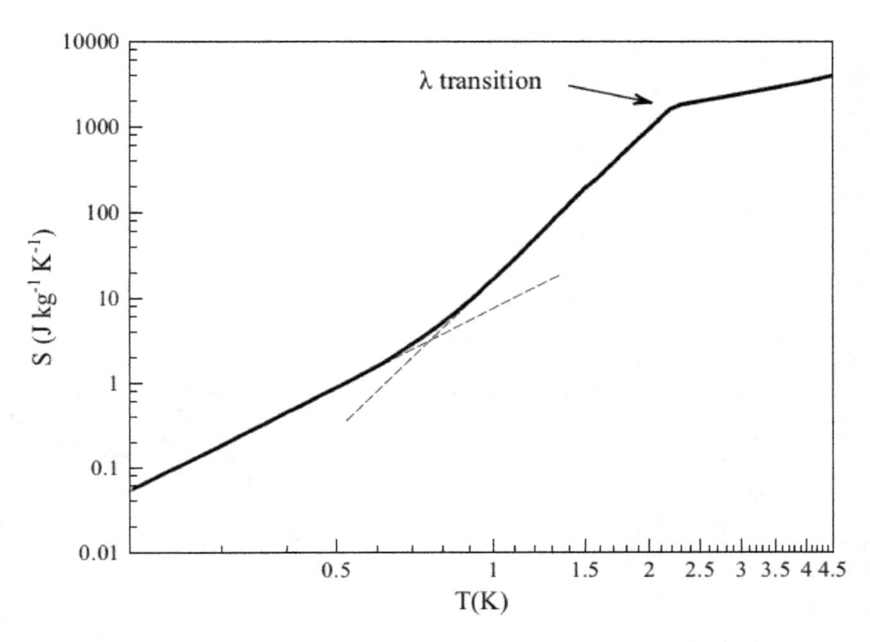

FIGURE 5.28: Entropy of liquid ^{4}He as a function of temperature. At the λ transition, the typical discontinuity of slope of a second-order phase transition is evident (adapted from [90]).

and specific heat of liquid ^{4}He is also important when understanding sub-K cryogenic systems (see Figs. 5.28 and 5.29).

Superfluid helium has the ability to flow through extremely narrow channels where normal liquid helium will not flow due to viscosity. These channels are referred to as *superleaks* meaning that only superfluid helium can pass.

Thermomechanical effect. It is an experimental evidence that if we establish a thermal gradient ΔT between two vessels containing liquid He-II — connected by a superleak — a difference in the level between the two liquids is established corresponding to a pressure difference ΔP.

Consider Fig. 5.31 in the initial configuration where the two reservoirs, connected by a superleak, are at the same temperature. Then, the liquid level will be identical. If we add a small amount of heat on the right reservoir ΔQ_R we will increase the temperature by a small amount ΔT. Superfluid helium will flow from the left reservoir to the right reservoir increasing the volume of the right reservoir by a quantity ΔV. The work done against gravity to raise the liquid level on the right reservoir is:

$$W = \Delta P \cdot \Delta V \tag{5.34}$$

At any temperature below T_λ, ^{4}He is a mixture of the normal and superfluid states. The fraction of superfluid depends only on the temperature:

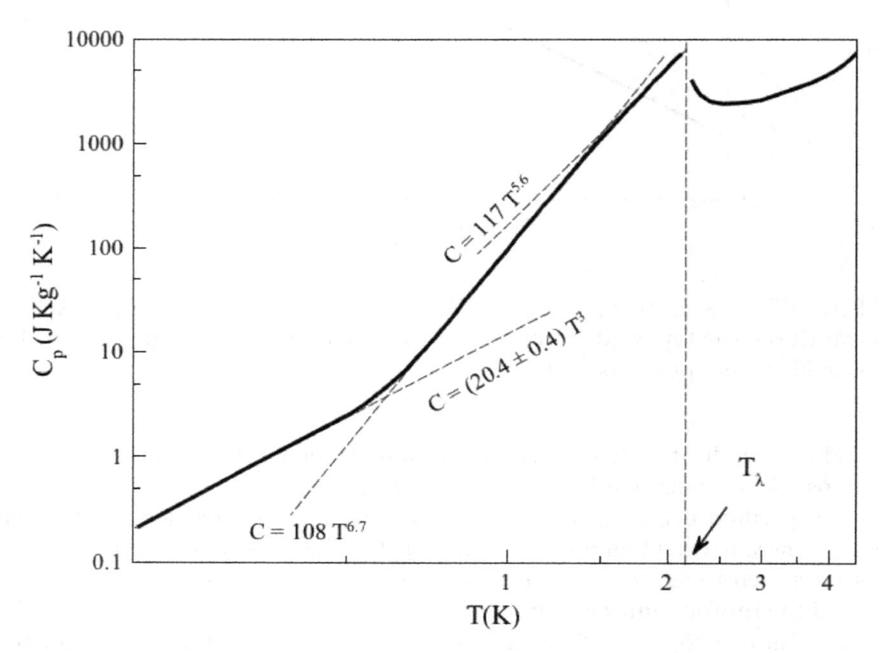

FIGURE 5.29: Specific heat of liquid ^{4}He as a function of temperature (adapted from [90])

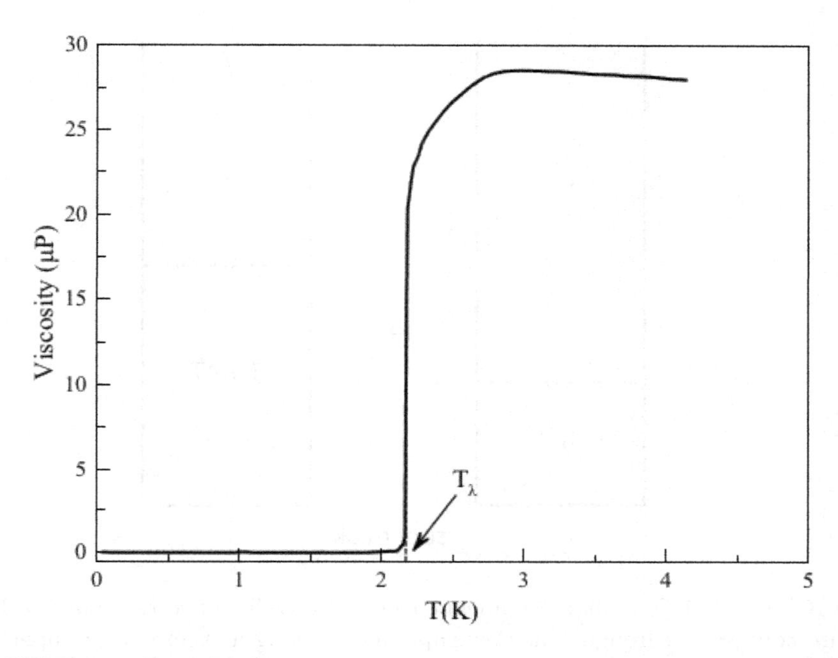

FIGURE 5.30: Liquid helium viscosity in cgs system versus temperature (data from [2, 4]

just below the transition temperature the liquid is practically all in the normal state, while just above absolute zero the liquid is practically all in the superfluid state (see Fig. 5.32).

Adding a small amount of heat on the right reservoir has generated the migration of pure superfluid helium from the left reservoir to the right reservoir. The left reservoir therefore has now a higher fractional content of normal fluid which would imply a higher temperature. In order to keep the temperatures constant, a correspondingly small amount of heat ΔQ_L must be removed from the left reservoir. Conservation of energy requires that:

$$W = \Delta Q_R - \Delta Q_L \tag{5.35}$$

In other terms, the work done against gravity to raise the level in the right reservoir is equal to the difference between the heat added to the right reservoir and the heat subtracted from the left reservoir. Using the Second Law of Thermodynamics, we have:

$$Q_L = msT \tag{5.36}$$

$$Q_R = ms(T + \Delta T) \tag{5.37}$$

where m is the mass transferred between reservoirs and s is the specific

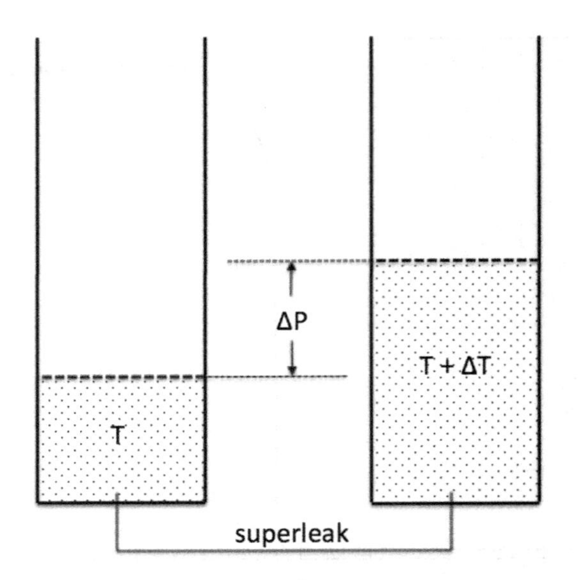

FIGURE 5.31: Thermomechanical effect in He-II. The two reservoirs of He-II are connected through a narrow capillary. As long as there is a temperature difference between the two reservoirs, then there is a differential pressure ΔP that maintains the level difference $\Delta P = \rho S \Delta T$ (adapted from [90]).

entropy of the normal fluid (since superfluid has zero entropy). We can also express the mass flow:

$$\dot{m} = \frac{\dot{Q}}{sT} \tag{5.38}$$

By noticing that $\rho = m/V$ and combining the above equations, we obtain a relationship between ΔP and ΔT, known as the London equation:

$$\Delta P = \rho s \Delta T \tag{5.39}$$

The London equation can also be written as:

$$\frac{dP}{dT} = \rho s \tag{5.40}$$

The London equation describes the so-called *fountain effect* in superfluid helium. There are many descriptions of the fountain effect, mainly involving superfluid helium creeping up vessels against the force of gravity following a positive temperature gradient. The London equation immediately tells us that, if a temperature gradient is present between two vessels, then a pressure difference is established that can even overcome gravity and have superfluid helium pouring out of its higher temperature containing vessel.

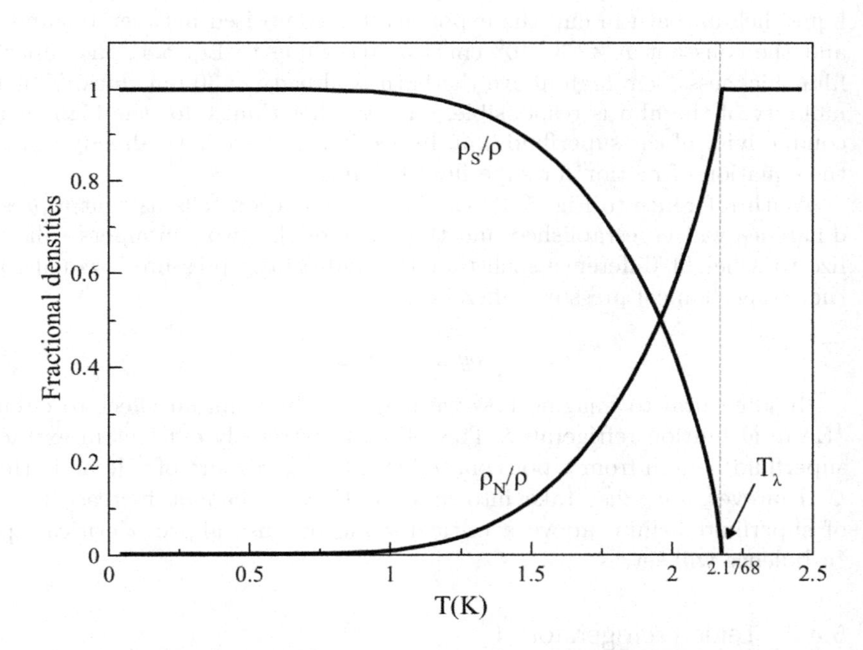

FIGURE 5.32: Fractional densities of normal and superfluid helium as a function of temperature

A related phenomenon is the so-called *superfluid film flow.* Normal liquids at saturation will stick to the surface of the containing vessel. Superfluid helium will also develop a film on the surface — usually less than 1 μm thick — sometimes called the *Rollin film.* The surprising fact is that this film is capable of flowing under the pressure difference due to the fountain effect. There is an empirical relationship between the height of the superfluid film above the liquid helium bath and the film thickness [90]:

$$d = \frac{K}{H^n} \tag{5.41}$$

where d is the superfluid film thickness in cm, H is the height above the liquid helium bath in cm, the exponent n is comprised between 0.3 and 0.45 and the constant $K \simeq 3 \cdot 10^6$ cm^{n+1}. According to Eq. 5.41 the superfluid film thickness 1 cm high above the bath is already $\sim$ 30 nm thick. This high mobility of the film is responsible, among other things, for the high thermal conductivity of the superfluid film. In Section 1.2.2 we have already discussed the equation of motion for superfluid helium.

With reference to Fig. 5.31, the London equation tells us that a pressure difference will be established and the levels of the two containers will stabilize to a height difference such that the hydrostatic pressure is equal to the thermomechanical pressure difference:

$$\rho g \Delta h = \rho s \Delta T \tag{5.42}$$

It is natural to imagine how we may use the fountain effect to circulate ^{4}He in a dilution refrigerator. This effect is extremely efficient in extracting superfluid helium from a pot containing a liquid mixture of ^{3}He and ^{4}He.

However, we must take into account that we cannot increase the flow of superfluid helium above a critical value because above a critical speed, turbulence will set.

5.4.2 Leiden refrigerator - I

The first ever dilution refrigerator where ^{4}He is circulated has been proposed by Taconis et al. in 1971 [82]. There are four important components: the mixing chamber, the de-mixing chamber, the connecting tube and the superleak. In a standard dilution refrigerator, the cooling action happens when ^{3}He atoms cross the phase separation surface from the concentrated phase to the diluted phase. Cooling can be achieved also if ^{3}He is mixed with superfluid ^{4}He — this process can be considered as ^{3}He *crossing* into ^{4}He generating cooling. Notice however that heat is generated when diluted ^{3}He is injected into concentrated ^{3}He (de-mixing). Fig. 5.33 shows a schematic diagram of a dilution refrigerator exploiting the circulation of ^{4}He. There are two pots, the mixing chamber and the de-mixing chamber connected with a tube. The mixing chamber is located above the de-mixing chamber because gravity is needed to operate this refrigerator.

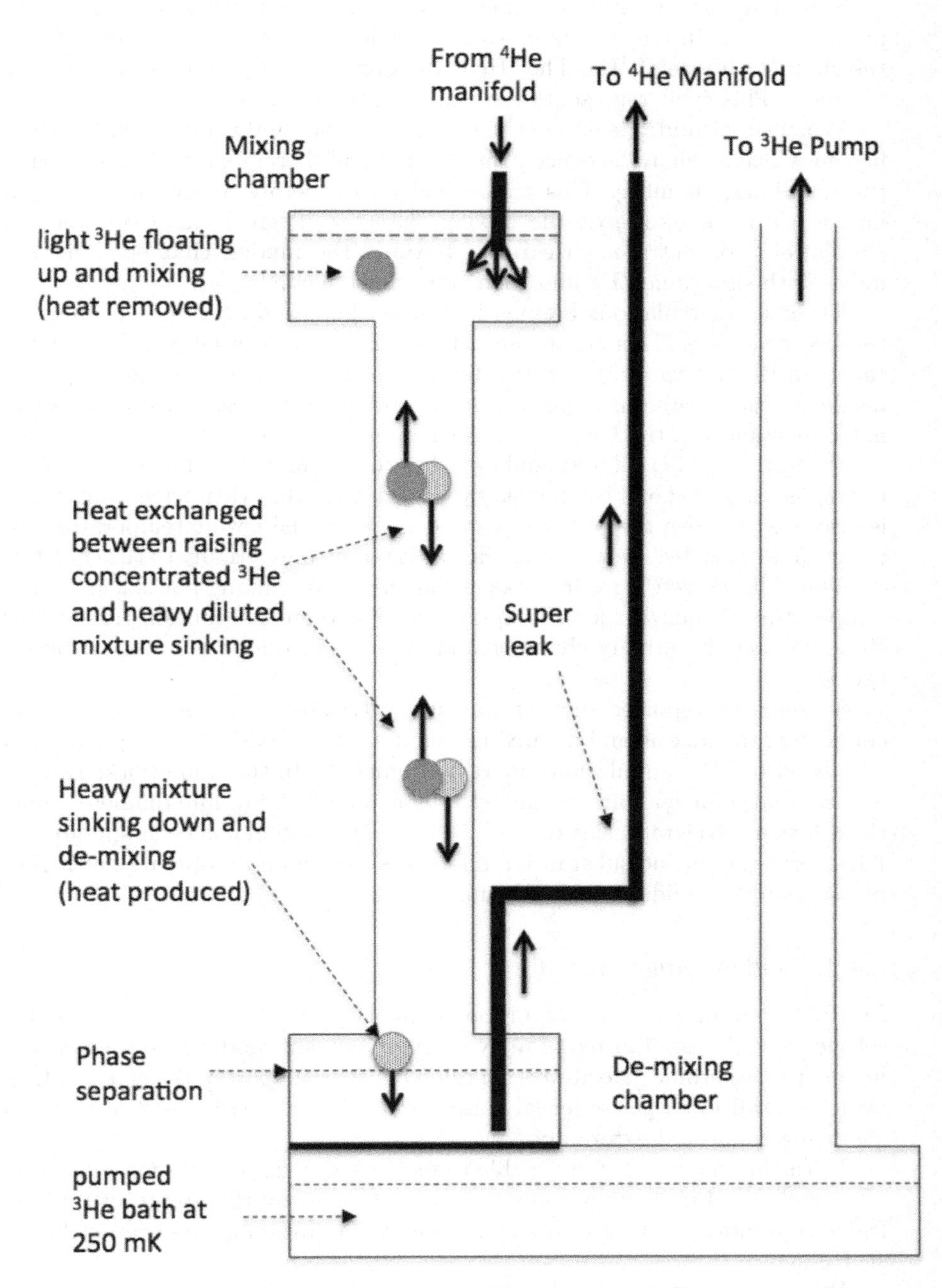

FIGURE 5.33: Schematics of the first dilution refrigerator circulating ^{4}He [82]

Superfluid ^{4}He is extracted from the de-mixing chamber by means of a superleak. The helium is circulated to an outside gas manifold to easily measure the circulation rate of ^{4}He. The ^{4}He is then returned superfluid to the mixing chamber. This cycle represents the main circulation.

When superfluid ^{4}He enters the mixing chamber and is mixed with ^{3}He, it becomes heavier than the concentrated phase and therefore it will sink towards the de-mixing chamber. This action will also generate a back flow of light liquid ^{3}He raising towards the mixing chamber. There is therefore another circulation: concentrated ^{3}He raises towards the mixing chamber where it mixes with superfluid ^{4}He and then sinks down again.

These two circulations happen in counterflow inside the tube connecting the two chambers. This action substitutes the need for a heat exchanger because, in this refrigerator, the two liquids are in maximal contact (they flow inside the same tube in opposite directions). Kapitza resistance is therefore not important and the heat exchanger is practically just a tube.

When the sinking diluted liquid reaches the de-mixing chamber, heat of de-mixing is generated and is removed by the action of the ^{3}He refrigerator, which is also used to keep the phase separated liquid at the lowest temperature. In the original work by Taconis, this ^{3}He gas is pumped externally to the cryostat.

This dilution refrigerator works as long as the de-mixing pot is kept at low temperature. When/if the ^{3}He liquid runs out, then the dilution refrigerator stops cooling the mixing chamber and the ^{3}He refrigerator needs to be re-cycled.

Taconis [82] reported that with a tube of 1.8 mm i.d. diameter 15 cm long connecting the mixing and de-mixing chambers, they reached a temperature of 60 mK with a ^{4}He circulation rate of 400 μmols^{-1}. In the same work, Taconis reports that a longer spiral wounded 100 cm long 2.75 i.d. mm diameter tube, the dilution refrigerator has reached 50 mK. The system has been operated at 1 Bar pressure but no substantial differences were found if operated at higher pressures up to solidification of helium.

5.4.3 Leiden refrigerator - II

In 1974, Pennings et al. [65] proposed an improved version of the Leiden refrigerator discussed in part I above. One important modification consists in having the superfluid ^{4}He always contained in the cold part of the cryostat thus avoiding additional gas external connections. There is still the ^{3}He pumped gas being connected externally.

With this version they were able to increase the ^{4}He circulation rate up to 10^{-3} mols^{-1}. Higher circulation rates were limited by the ability of the ^{3}He fridge to remove the heat of de-mixing due to limitation in the heat exchange surfaces.

What is important, in this version, is that a lot of engineering tests have been performed to optimize the geometry of the tube connecting the mixing and the de-mixing chamber. We already mentioned that this tube acts also

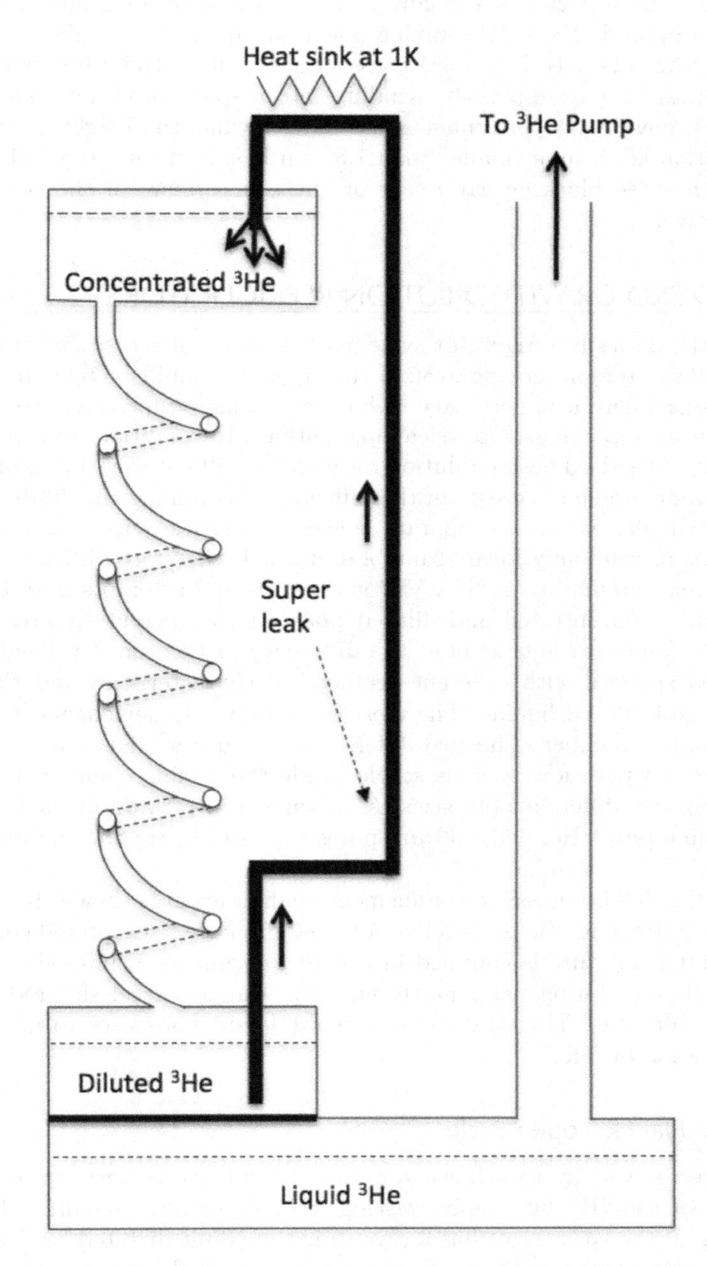

FIGURE 5.34: Improved version of the Leiden refrigerator where the ^4He is circulated cold (adapted from [65])

as an almost perfect heat exchanger because the liquids exchanging heat (^{3}He floating up and ^{3}He + ^{4}He sinking down) are in the same tube. Winding the connecting tube (see Fig. 5.34) is essential for the optimal functioning of the refrigerator: for example, the winding into a spiral of 60 cm long tube, 2.8 mm i.d. gave good performances no matter what the height of the winded up section is. A tube smaller than 1.8 mm has been reported to be too narrow otherwise blockage can occur or the counterflows of the two liquids is ineffective.

5.5 ZERO GRAVITY DILUTION REFRIGERATORS

In all the dilution refrigerator systems that we discussed so far, gravity plays an essential role in moving around the cryogenic liquids. There are space applications where it is necessary to have cryogenic temperature around $\sim$ 100 mK, for example in astrophysical applications. In conditions of zero or microgravity, a standard design dilution refrigerator will not work for many reasons. The major problem consists in the difficulty of confining the diluted and concentrated phases in two separate spaces. In zero gravity, phase separation happens in randomly located bubbles and it is therefore difficult to produce a coherent circulation of ^{3}He. Various designs and attempts have been made to confine concentrated and diluted phases into different sintered materials [70, 71]. The idea is to exploit the difference in the van der Waals between sintered sponges with different geometrical characteristics and the concentrated and diluted liquids. The most important confinement must happen in the mixing chamber. The two different sintered materials must ensure that the phase separation remains stable inside the mixing chamber for example by using two different pore sizes. As discussed in [71], 80 μm pores are used to confine pure ^{3}He, while 40 μm pores are used to confine the mixture ^{3}He + ^{4}He.

In Fig. 5.35, a possible confinement configuration is shown. In the mixing chamber the pure ^{3}He is confined in the 80 μm pore size sintered copper while the diluted mixture is confined in the 40 μm pore size sintered copper. The pure ^{3}He circulation happens through the stainless steel sintered rods with 80 μm pore size. The still also has two different pore sizes to maximize the evaporation of ^{3}He.

5.5.1 Planck's open cycle

An alternative way to achieve zero gravity dilution consists in avoiding the return of the ^{3}He by simply wasting it in space after having achieved the cooling. If the space craft has a cryogenic time limit dictated by, for example, the duration of the main liquid helium bath or just the mission duration, then it is possible to conceive a dilution system where the ^{3}He is not circulated providing that the spacecraft has a big enough reservoir of it. If we force pure

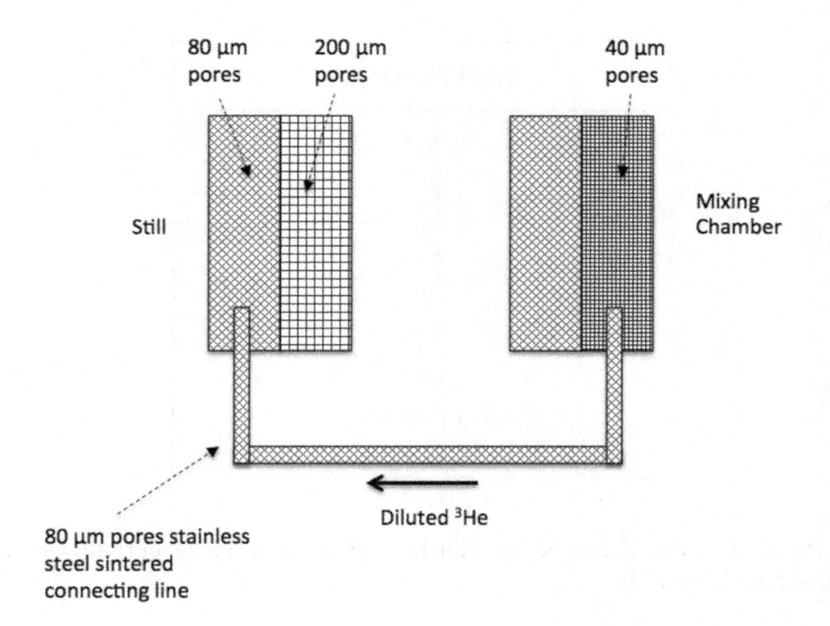

FIGURE 5.35: Zero-gravity dilution refrigerator using various sintered materials for liquids confinement (adapted from [71])

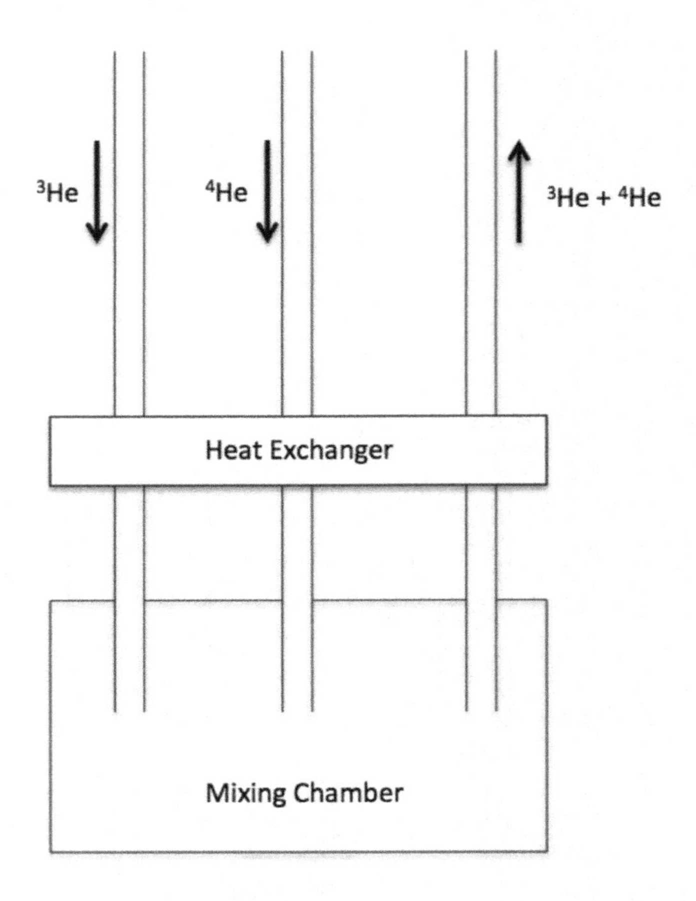

FIGURE 5.36: Zero-gravity dilution refrigerator using enthalpy of mixing (adapted from [8])

liquid ^{3}He to diffuse into diluted mixture, for example through a simple Y junction, then dilution cooling can be achieved independent from gravity.

Although this design has the clear disadvantage of being an open cycle, thus wasting the precious ^{3}He, it does not require a still and can work in any orientation. In Fig. 5.36, a schematics of the cold part of this refrigerator is shown. The system needs to have the two gases, ^{3}He and ^{4}He, precooled before entering the mixing chamber. It is important to note that this refrigerator does not use the osmotic pressure to extract the ^{3}He from the mixing chamber. Instead, if the velocity of liquid ^{4}He is high enough, it can force the mixture containing ^{3}He out of the mixing chamber. This mechanism works if and only if the friction between the liquid ^{4}He entering the mixing chambers and the diluted ^{3}He is high enough. Kuerten et al. [44] have shown that such a mutual interaction exists and is enough to properly drive the diluted phase out of the

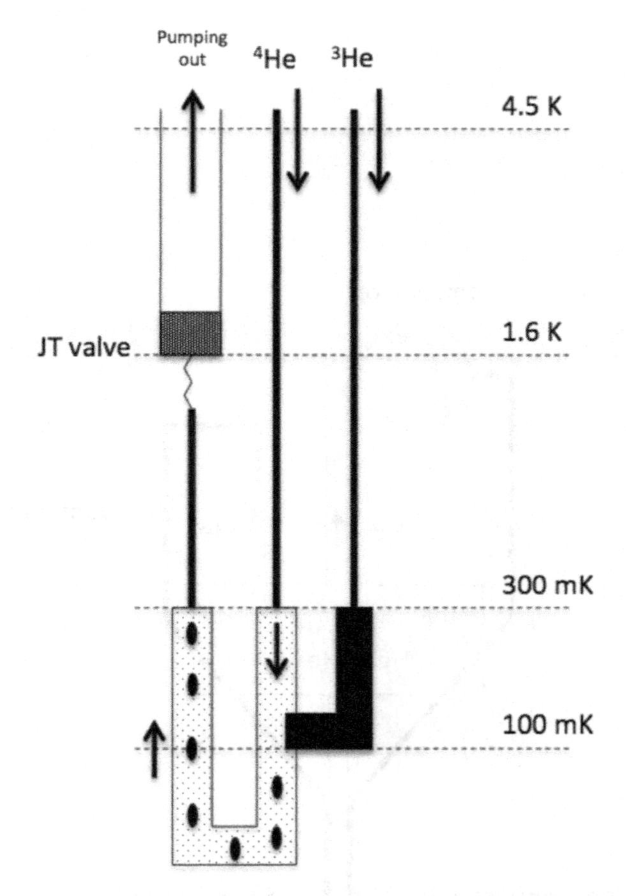

FIGURE 5.37: Schematic of the Planck satellite dilution stage (adapted from [89])

mixing chamber. With a simple continuous heat exchanger, temperatures as low as 53 mK have been reported with ^{3}He flow rate of 4 μmols^{-1} and ^{4}He flow rate of 12 μ mols^{-1} for an early prototype.

The final space version [89] is schematically shown in Fig. 5.37.

5.5.2 Closed cycle

The Planck refrigerator of the previous section has been successfully deployed and operated seamlessly for the whole duration of the Planck satellite mission. It is possible to conceive a way to return the ^{3}He in a Planck-like system. Such a possibility is explored by Chaudhry et al. [13] where a closed-cycle adaptation of the Planck dilution cooler is presented.

As shown in Fig. 5.38, the circulation of ^{3}He is achieved by means of an

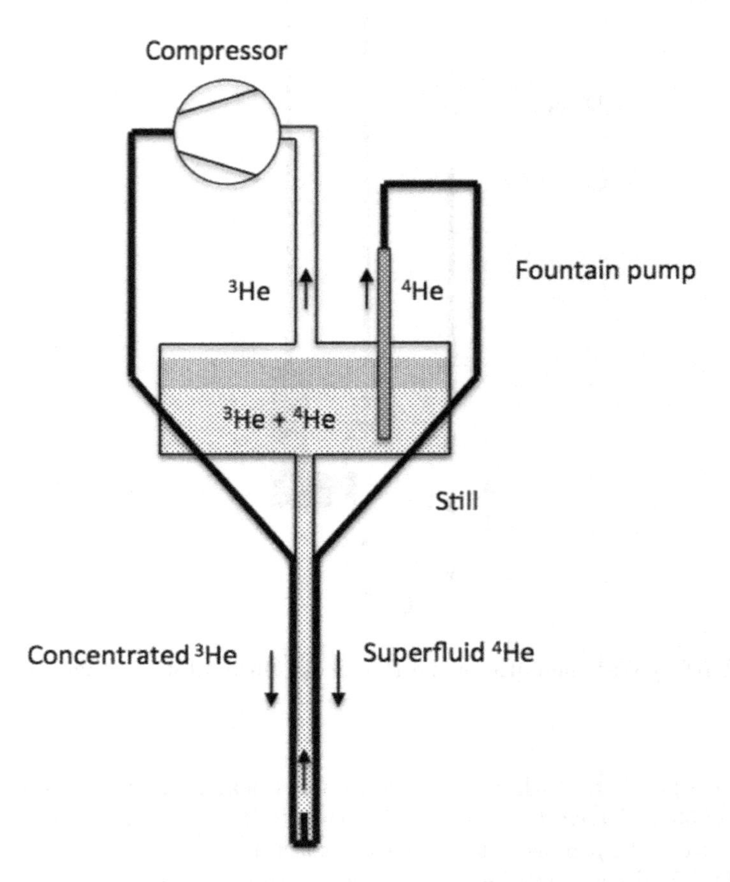

FIGURE 5.38: Schematic of the Planck-like closed-cycle zero gravity dilution refrigerator (adapted from [13])

external compressor, while the circulation of ^{4}He is achieved by a superfluid fountain pump. The returning liquid ^{3}He and ^{4}He is pre-cooled to 1.7 K before being thermalized in the still at $\sim$ 1 K. The liquids then enter the counterflow heat exchanger and then are mixed in the bottom Y-junction (mixing chamber). Friction between liquid ^{3}He and ^{4}He plus the pressures generated by the compressor and the fountain pump push the mixture towards the still where ^{3}He and ^{4}He are separated.

This design, although very attractive for the absence of cold valves, still needs a compressor with its associated power consumption, mechanical moving parts and vibrations.

III

Simulations

CHAPTER 6

Thermal Modeling

THIS CHAPTER will cover the thermal modeling of an experiment. Having a detailed thermal model is important in designing a cryostat and a sorption fridge, because it gives information about the cooling power required by each component. Thermal modeling is also useful for estimating the cooldown time for the cryostat, which is of increasing importance for larger experiments.

6.1 COOLDOWN MODELING APPROACH

Estimating the cooldown for a cryostat is crucial if the experiment needs to be operative quickly. To cool down from room to cryogenic temperature, it is possible to use either a mechanical cooler or liquid cryogens. The first solution is typical of any *dry dewar* and has the advantage that no cryogens have been used to cool down the cryostat and hence no continual top-up is needed.

Many mechanical coolers are available commercially; most common are Gifford-MacMahon and pulse-tube coolers, where through a process of expansion and compression of a recirculating helium gas, it is possible to reach temperatures of 4 K or below. The pulse-tube is connected to the mass that needs to be cooled through some heat straps and will remain connected after the cooldown to keep the mass cooled.

The second solution of using liquid cryogens is typical of a *wet dewar*. In this case, it is possible to fill a tank with liquid nitrogen (boiling at 77 K), so that the mass in contact will start to cool. Once the cold mass reaches an equilibrium temperature with the nitrogen vessel, a second vessel (typically concentric with the nitrogen stage) may be filled with liquid helium (boiling at 4.2 K) to reach a similar temperature to the one provided by a mechanical cooler.

In order to precool a cryostat, it is also possible to use both mechanical coolers and liquid cryogens in combination. In particular, it has been common to use a system to pump liquid nitrogen through a precooling circuit in a large cryostat to reach 77 K and then use a mechanical cooler to reach and maintain the base temperature.

In order to estimate the cooling time, it is possible to consider that the energy that needs to be removed is:

$$Q = m \int_{T_0}^{T_1} C(T)dT \tag{6.1}$$

where $C(T)$ is the heat capacity of the mass that needs to be cooled. It is possible to derive Eq. 6.1 with respect to time as:

$$mC(T)\frac{dT}{dt} = \frac{dQ}{dt} = \dot{Q} \tag{6.2}$$

where the infinitesimal version of Eq. 6.1 has been considered, and $\dot{Q}$ is the cooling power given by the mechanical cooler or the liquid cryogen boil off. The cooling power of the liquid nitrogen is given by:

$$\dot{Q} = \dot{n}\lambda(T) + \dot{n} \int_{T_0}^{T_1} C(T)dT \tag{6.3}$$

where $\dot{n}$ is the evaporation rate in case of a wet cryostat or the flow rate in case of the plumbing lines. Eq. 6.3 describes the ideal cooling power of the liquid nitrogen; however, in reality, this is lower due to the fact that not all of the enthalpy (second term of the equation) is used to cool down the mass. Moreover, in the case of the plumbing system, the efficiency of the heat exchanger must also be considered.

The cooling power and number of stages of a mechanical cooler is dependent on the model chosen. For an example of the two-stage cooling power of a pulse tube cooler from room temperature, the reader is directed to [31].

Mechanical coolers are typically linked to the cold mass by a heat strap, the conductance of which will give the effective cooling power seen by the mass. In this case, the mass will see a cooling power which is given by:

$$\dot{Q} = \frac{T_{HS} - T_m}{R_K} \tag{6.4}$$

where R_k is the thermal contact resistance between the heat strap and the mass. This modeling method is usually called *lumped*, because it replaces the continuous geometry of the system with a series of elements which approximate the behaviour of the system by considering the equivalent circuit. By analogy with electrical circuits, one may consider that electrical resistance is equivalent to thermal resistance and capacitance is equivalent to the heat capacity of the mass. For example, in case of a mechanical cooler connected to a mass, the equivalent circuit is presented in Fig. 6.1.

Until now, the mass has been considered small enough to be isothermal during the cooldown process. However, sometimes the mass that needs to be cooled is large enough or the thermal conductivity of the material is not high

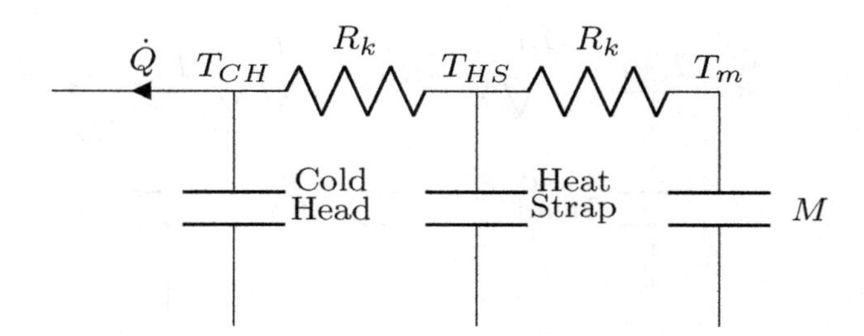

FIGURE 6.1: Equivalent circuit for cooling with a mechanical cooler

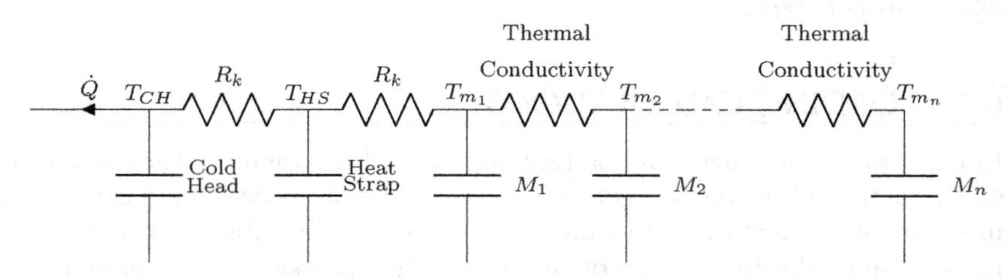

FIGURE 6.2: Equivalent circuit for cooling with a mechanical cooler when the mass is divided into submasses

enough that there will be a temperature gradient across it. In this case, the simplest approach is simply dividing the mass into submasses that can be considered isothermal, where the resistance between each element is proportional to the thermal conductivity of the material. An equivalent circuit for this situation is presented in Fig. 6.2.

The use of mechanical coolers is limited to a base temperature of around 3 K, while the use of liquid nitrogen is limited to 77 K.

Often, components with a base temperature requirement below 3 K are thermally well-isolated from the warmer stages, and hence cannot be precooled by these stages directly in a reasonable time. Instead, radiative cooling may be used by selecting materials with a suitable emissivity as described in Chapter 2. Another possible solution is the use of heat switches that connect the warmer stages to the colder stages, also described in Chapter 2.

In the cooldown model described so far, the heat load on the different stages from warmer stages has not been considered. This may be reasonable at higher temperatures, where the cooling power is much greater than the loading; however, at lower temperatures it is important to include these additional loads in the model. In this case, the equivalent circuit for the easiest case is presented in Fig. 6.3.

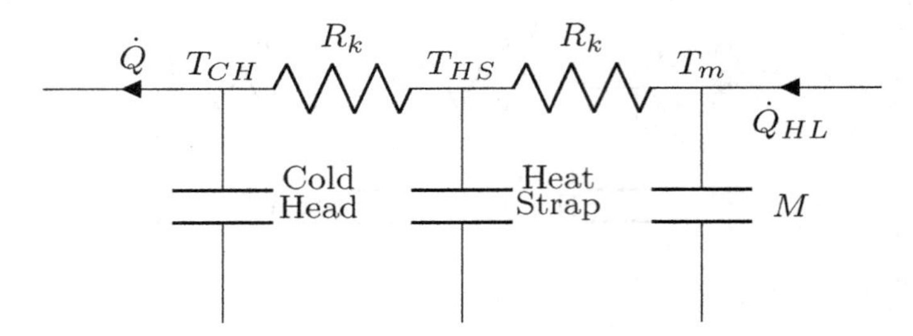

FIGURE 6.3: Equivalent circuit for cooling with a mechanical cooler when a heat load is present

6.2 LOADING CALCULATIONS

In the absence of convective loading due to the high vacuum, the main sources of thermal loading on the cold stages of a cryostat are conduction through mechanical connections and radiative loading, as described in Chapter 2. We now consider the calculation of these loads for inclusion in the model.

6.2.1 Conduction

Conduction on each stage is given principally by (a) conduction through the supports of each stage and (b) conduction through the wires to the various thermometers, heaters and other electrical components.

In order to compute the contributions of these two components, Eq. 2.5 may be used. This formula describes the heat conduction in one dimension, which is appropriate for our purposes. Indeed, wires usually have a very small radius, compared to the length, and the different supports usually have a transverse temperature gradient which is usually negligible with respect to the on-axis temperature gradient.

Conductive loading through the mechanical supports is usually minimized using low thermal conductivity material such as fiberglass or carbon fiber. These materials have great mechanical strength (it is necessary to remember that these may need to support tens or hundreds of kg) and a relatively low thermal conduction. In Fig. 6.4, the ratio between elastic modulus and thermal conductivity at low temperature is presented. It is evident that, for all the materials shown, the ratio increases at low temperature due to the decreasing thermal conductivity.

Additionally, wires need to have a relatively low thermal conductivity, as well as being well-thermalized on each stage. For example, a wire from the room temperature stage to the 4 K stage needs to be *heat sunk* on each stage in between; by intercepting the heat load in this way, the loading on the cold stages is reduced. This is shown by Eq. 2.5 where the limits of the integral are given by the temperatures of the stages that the section of wire is being sunk at. It is clear that if the wiring is not sunk at each stage, then the upper

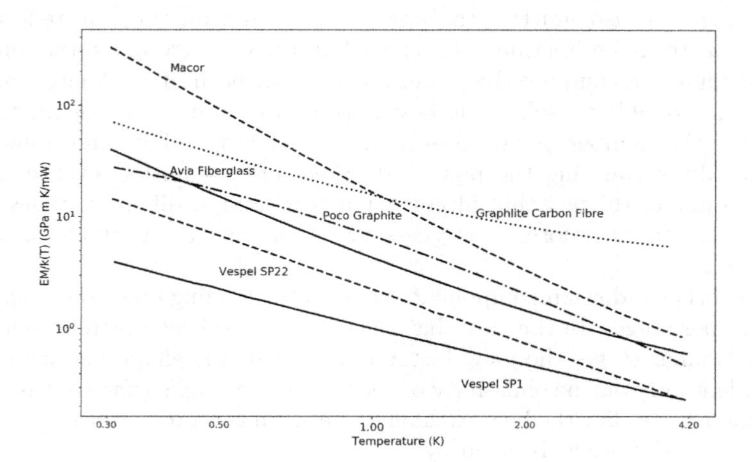

FIGURE 6.4: Ratio between elastic modulus and thermal conductivity at low temperature (data from [72])

limit on the integral will be much higher and hence the load on the cold stages much greater.

6.2.2 Radiation

Radiative loading on the cold stages results from (a) radiation from the previous hotter stage and (b) in the case of an optical cryostat, radiation from the apertures between the stages.

Apertures between stages are necessary to collect radiation coming from outside the cryostat whilst keeping the detectors cold. Usually on the external shell there is a so-called *window* that is able to bear the differential pressure between the external environment and the internal vacuum. After this window a series of *filters* at different stages are placed. These filters are useful to reduce the incoming radiation on each stage to the only band (or bands) scientifically important.

In order to characterize the filters, it is important to define other parameters than the only emissivity introduced in Subsection 2.1.3. In particular, we need to introduce the *reflectivity*, ρ, *absorbivity*, α, and *transmissivity*, τ. These describe the fraction of radiation reflected, absorbed, and transmitted, respectively. For a grey body, the emissivity is equal to the absorbivity, so $\epsilon = \alpha$. Moreover, in this case a useful relation is:

$$1 = \alpha + \tau + \rho = \epsilon + \tau + \rho \tag{6.5}$$

The total load on each stage is given by the fraction of the energy absorbed, so it is proportional to α, or, equivalently, ϵ. However, each filter is not only absorbing, but also emitting (as a grey body), transmitting and reflecting, so there is a strong interaction between each filter. This strong interaction makes it challenging to compute the power absorbed by each filter. Analytically, it is almost impossible to do so; the best approach is to use ray-tracing and then compute the number of rays absorbed on each stage. The main difficulty in analytically estimating the power adsorbed and re-emitted by the different filters is due to the fact that filters usually are used at different stages so they can be distant; therefore, it is necessary to compute the view factor between the filters.

The other radiation component is due to the loading on colder stages from the warmer stages. In the case that the colder stage is completely surrounded by the wamer stage, the view factor is 1, and so the shape factor defined is dependent only on the emissivity of the two interacting surfaces. It is possible to demonstrate that the heat exchanged between two grey bodies, one of them enclosed by the other, is given by:

$$P = \frac{\sigma\left(T_1^4 - T_2^4\right)}{\frac{1}{\epsilon_1} + \frac{1}{\epsilon_2} - 1} \tag{6.6}$$

This contribution can be extremely high, especially at high temperatures, so the easiest way to reduce the radiative load consists in using multiple layers of material between the stages. In this case the heat exchanged is given by:

$$P = \frac{\sigma\left(T_1^4 - T_2^4\right)}{\frac{1}{\epsilon_1} + \frac{1}{\epsilon_2} - 1 + N\left(\frac{1}{\epsilon_i} + \frac{1}{\epsilon_e} - 1\right)} \tag{6.7}$$

where N is the number of layers, ϵ_i is the emissivity of the internal face of each layer, and ϵ_e is the emissivity of the external one. A common material used is *multi-layer insulation*, which is composed of N layers of thin sheets. Usually these sheets are made by aluminized mylar.

If the geometry of the cryostat is more complicated, such as where a 4 K stage is composed of multiple components, then each element of the stage may have a direct line of sight to other elements of the same stage. This means that the component will not be completely irradiated by the hotter stage on all sides; therefore, it is necessary to compute the view factor between the different bodies. To do so, it is possible to use ray-tracing codes, as in the case of the filters.

6.3 FULL CRYOSTAT MODEL

Using the modeling approach described above, it is possible to construct a model in Python (or another programming language of the reader's choice) to simulate the transient thermal behaviour of a cryostat. Example code for a medium-sized pulse-tube cooled cryostat is given in the Appendix in Section B.3.

It should be noted here that there are several commercial packages that may be used for thermal simulations. However, it has been the authors' experience that homemade codes, whilst not without their challenges, are inherently more flexible and customisable to the user's needs; an example of this is allowing cooling powers to be applied which are a function of the temperature at multiple points in the system, such as in the case of a two-stage mechanical cryocooler.

The model should firstly define, for each of the lumped elements, the initial temperature and heat capacity as a function of temperature.

It should then define, for each coupling between elements, the parameters for radiative and conductive heat transfer (described in Section 6.2). Here, we may define a cooling power applied to an element which is a function of the temperature of two elements, as would be the case for a two-stage cryocooler (in the code presented in the Appendix, this is done by interpolating values from the capacity maps).

The model has a defined timestep. It loops so as to calculate the heat transferred in each coupling during said timestep, finding the net heat transfer for each element, calculating the new temperature from the (temperature-dependent) heat capacity, updating all of the temperatures, and then moving on in the timestep. In our code, the loop terminates when the defined end time is reached, although it could just as easily be set to terminate when some other condition is met (for example, a steady state condition on one or more elements).

The simulated cooldown for the cryostat described above is given in Fig. 6.5 below and is in good agreement with experiment, showing a cooldown time from room temperature of $\sim$ 4.5 hours.

There are, however, limitations to such a model. Firstly, the model approximates continuum behaviour by discretising in both time and space. The model should have a suitably small timestep and suitably fine geometric discretisation such that the solution given is suitably accurate. In practice, this may be done by reducing the timestep and increasing the number of elements until the solution does not change (relative to the level of precision required in the result). Furthermore, it has been the authors' experience that the effects of thermal boundary resistance between components (e.g., that of a bolted joint), which become significant at very low temperatures, may be difficult to capture as they are strongly dependent on the torque applied to the bolt, the material finish, presence of interstitial material, etc. As such, great care must be taken when interpreting these types of results; any thermal model should always be validated with respect to experimental data and should be used as a guide in design only.

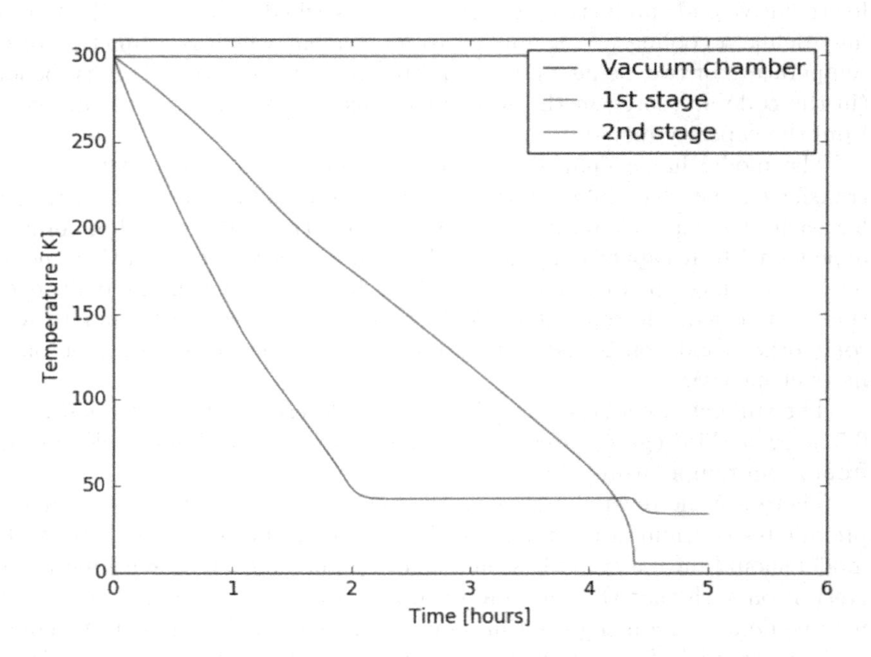

FIGURE 6.5: Simulated cooldown of a medium-sized pulse-tube cooled cryostat

FURTHER READING

Cook, R.D., Malkus, D.S., Plesha, M.E. and Witt, R.J. (2001). *Concepts and Applications of Finite Element Analysis.* John Wiley & Sons.

Hughes, T. (2012). *The Finite Element Method: Linear Static and Dynamic Finite Element Analysis.* Courier Corporation.

Szabo, B.A. and Babuska, I. (1991). *Finite Element Analysis.* John Wiley & Sons.

APPENDIX A

Helium Properties

In the main chapters, and especially in Chapter 1, we have introduced several helium properties for both isotopes. In this Appendix, we report some tabular data that can be useful to design a helium sorption cooler. More ^{4}He properties can be found in [21]. Also, for ^{3}He it is possible to find many properties using the equation of state as shown in [37, 38, 39, 40].

A.1 LATENT HEAT

T(K)	$\lambda(T)$	T(K)	$\lambda(T)$	T(K)	$\lambda(T)$
0.101123883	21.35080248	1.371663271	42.63862945	2.650506837	40.86181289
0.146796868	22.30699379	1.417336255	43.10073635	2.696179821	40.03930035
0.190393808	23.21634413	1.46300924	43.55493791	2.741852806	39.12587626
0.233990748	24.13280927	1.508682224	43.96961272	2.78752579	38.14130403
0.279663732	25.06133186	1.554355209	44.38428752	2.829046685	37.19050165
0.323260672	25.8951861	1.600028193	44.74757755	2.866415491	36.2835704
0.366857612	26.79228313	1.645701178	45.05553012	2.901708252	35.30436929
0.412530596	27.66546827	1.691374162	45.36348269	2.932848923	34.36972947
0.458203581	28.4912213	1.737047147	45.62795584	2.961913549	33.41665146
0.503876565	29.35254841	1.782720131	45.82918618	2.988902131	32.46273408
0.54954955	30.23759157	1.828393116	46.01065314	3.013814668	31.57630214
0.595222534	31.094966	1.8740661	46.14073533	3.038727205	30.6391442
0.640895519	31.94048241	1.919739085	46.21943274	3.062156853	29.61206163
0.686568503	32.77414079	1.965412069	46.28231945	3.086476234	28.49931657
0.732241488	33.56431974	2.011085054	46.26615265	3.109312727	27.38022399
0.777914472	34.30311392	2.056758038	46.21441177	3.130073174	26.35215096
0.823587457	35.01819204	2.102431023	46.11128612	3.150833622	25.32842588
0.869260441	35.72931749	2.148104007	46.01211314	3.168480002	24.24086242
0.914933426	36.42463224	2.193776992	45.81412328	3.189523547	22.96463349
0.96060641	37.13575769	2.239449976	45.48569514	3.208962875	21.79132546
1.006279395	37.80735639	2.285122961	45.21655712	3.225571233	20.59407437
1.051952379	38.49476578	2.330795945	44.8683656	3.241883013	19.50166774
1.097625364	39.15055378	2.37646893	44.48460001	3.254635859	18.50794318
1.143298348	39.77472037	2.422141914	44.03363893	3.265016083	17.55657609
1.188971333	40.41469767	2.467814899	43.55896181	3.276088322	16.55840333
1.234644317	41.03095891	2.513487883	42.99337315	3.28577653	15.44151736
1.280317302	41.61164607	2.559160868	42.35268366	3.298648008	14.16722543
1.325990286	42.11327973	2.604833852	41.68037277	3.300308844	13.61401042

TABLE A.1: Latent heat of ^3He

T(K)	$\lambda(T)$	T(K)	$\lambda(T)$	T(K)	$\lambda(T)$
0.97596783	83.98394089	1.988057846	91.11548957	3.597544108	90.74813641
0.998319591	84.15314401	2.033731913	91.1147684	3.63099927	90.58933162
1.019184822	84.32916254	2.079558956	91.01216565	3.662381655	90.42835043
1.036506782	84.48597007	2.115116949	90.83552085	3.691679555	90.27298992
1.055242203	84.64951258	2.130360456	90.6691201	3.71889772	90.12008794
1.080943973	84.81520619	2.150402806	90.49445685	3.746132534	89.95609681
1.096289859	84.96468987	2.163629571	90.32696792	3.771904197	89.77896072
1.117513731	85.11693093	2.176308179	90.17888948	3.798535956	89.62156314
1.140320846	85.26817672	2.187832738	90.00932474	3.823319929	89.43065778
1.159947271	85.43087349	2.203808047	89.83837291	3.855206765	89.21024085
1.178396097	85.58776719	2.21236374	89.67086491	3.874385017	89.03496461
1.196840553	85.74757179	2.231337418	89.47820776	3.898782488	88.84728638
1.215291347	85.90315559	2.268931506	89.32816934	3.920140428	88.67281369
1.233515785	86.05586446	2.314645307	89.30098543	3.941377777	88.50882485
1.25634328	86.21548435	2.36019322	89.38428348	3.962405331	88.33093201
1.275352666	86.3761306	2.401470405	89.5465943	3.986781952	88.13479358
1.297348633	86.55883502	2.436509043	89.71583966	4.006553742	87.96365186
1.319924297	86.73254639	2.467433478	89.85985346	4.025495641	87.7921594
1.338360504	86.89784532	2.50248731	90.01897998	4.046552698	87.59461797
1.358860019	87.07162586	2.539621255	90.17539672	4.063720278	87.4196969
1.380371358	87.26285542	2.576759025	90.32926643	4.081900929	87.2504198
1.404879738	87.45521008	2.618039422	90.48943817	4.099240411	87.0707457
1.428574787	87.65154821	2.661408628	90.64110905	4.117969116	86.88761327
1.453243622	87.81385349	2.706857205	90.79056398	4.132972841	86.72728054
1.474672918	87.98290946	2.752303796	90.94134204	4.151684396	86.55556936
1.495442183	88.15261115	2.797769261	91.0795503	4.16671327	86.37848739
1.515369563	88.31239233	2.843258566	91.20188091	4.181475131	86.22560882
1.534695969	88.47738093	2.888768731	91.31031857	4.194470491	86.06409495
1.557377491	88.62997149	2.934299757	91.4048633	4.210749877	85.88814833
1.581518912	88.79787286	2.979864556	91.47691468	4.225846631	85.70975137
1.601466412	88.95522804	3.025445249	91.53838097	4.242729919	85.52664808
1.623282743	89.12988009	3.071043822	91.58793902	4.257518446	85.35600955
1.646181591	89.28587186	3.116681135	91.6116959	4.270222685	85.19086173
1.668778599	89.44536833	3.162327389	91.62949865	4.282934207	85.02086242
1.696555006	89.61199945	3.208001456	91.62877748	4.295640267	84.85450173
1.718105408	89.77721282	3.253705324	91.60820925	4.304886169	84.68819568
1.741737367	89.93875479	3.299438992	91.56779396	4.318267704	84.49095185
1.763304614	90.09274906	3.34519948	91.50951631	4.329404111	84.29259363
1.787986458	90.24639094	3.390982816	91.43602259	4.34445306	84.14603466
1.812657374	90.40731008	3.436811845	91.3320967	4.351243683	83.96893838
1.839402564	90.56940931	3.482677628	91.20369277	4.364528343	83.77276717
1.870294023	90.73538483	3.524415064	91.05947675	4.384068211	83.57173244
1.903280697	90.88859237	3.562020289	90.90202142	4.403434635	83.39403562

TABLE A.2: Latent heat of ^{4}He

A.2 VAPOUR PRESSURE

T(K)	P (mbar)	T(K)	P(mbar)	T(K)	P(mbar)
0.567585433	0.000167227	0.9189942	0.047221746	1.755099848	9.902136431
0.575852487	0.000200392	0.935936984	0.057357586	1.793938649	11.58356871
0.584658496	0.000243217	0.95674157	0.069526928	1.83929653	13.63136888
0.592355822	0.000289521	0.974612139	0.083353657	1.891704909	16.39787722
0.600150218	0.000343745	0.992033107	0.09909382	1.942495699	19.24556437
0.609535458	0.000417227	1.013457133	0.11986072	2.000827634	22.80931877
0.618302237	0.000498209	1.030581578	0.140357909	2.064090313	27.11816541
0.626999132	0.00060105	1.046574873	0.165754455	2.126097149	32.26972839
0.634396726	0.000706011	1.067945912	0.199520776	2.189935045	38.19751889
0.643162212	0.000843589	1.08583188	0.231418356	2.252138714	44.51371523
0.652526716	0.001008998	1.107446411	0.271930778	2.3233144	52.64862793
0.66112616	0.001206812	1.129592831	0.330215591	2.404158271	62.84317248
0.670751751	0.001443073	1.148279352	0.385397291	2.483897336	73.83355164
0.682169617	0.001751329	1.170136933	0.456454547	2.562351132	86.7520421
0.693171178	0.002102988	1.195634867	0.544187593	2.647333971	101.8811028
0.703386372	0.002518966	1.21888103	0.64843661	2.739311518	119.3402409
0.715195033	0.003076567	1.242060789	0.76309745	2.834316986	136.80151
0.728074298	0.003788241	1.267628295	0.899448128	2.932734997	159.1312082
0.739212412	0.004607617	1.295757505	1.075495739	3.034553073	184.7188381
0.751510946	0.005514433	1.324467218	1.270590961	3.13984612	212.9306658
0.764541361	0.006620642	1.354869131	1.508445008	3.248805031	245.7938413
0.773863978	0.007815353	1.38172359	1.789358325	3.36160919	285.7147001
0.789225288	0.009369092	1.414567296	2.148542391	3.478283637	330.5019388
0.799881848	0.011219533	1.445900769	2.532731108	3.598959528	380.4479732
0.815293607	0.013451622	1.480237786	3.017067455	3.723815053	437.6366224
0.828865561	0.016153256	1.517724124	3.602152308	3.852980019	502.3696558
0.843221594	0.019262722	1.553776111	4.300410538	3.986435013	566.7103326
0.857157335	0.022920507	1.590598009	5.033048092	4.12457544	642.8681444
0.87403398	0.027994924	1.630819252	5.928948588	4.267551664	732.3184288
0.887756355	0.033113256	1.672127772	7.092309138	4.415441947	831.3111752
0.901892531	0.039324382	1.714464636	8.451460927	4.547020538	908.1260555

TABLE A.3: Vapour pressure of ^4He

T(K)	P (mbar)	T(K)	P(mbar)	T(K)	P(mbar)
0.229954921	0.000122586	0.422264756	0.049299526	1.126652647	15.49297603
0.233622927	0.000145208	0.431624634	0.058434277	1.165776436	18.03442079
0.237272755	0.000175274	0.441191982	0.069261614	1.206245012	20.90510079
0.239982292	0.000203344	0.450967342	0.08182569	1.248104115	24.13154069
0.244132636	0.000244567	0.460957908	0.096563009	1.291403489	27.75897548
0.247679908	0.000289378	0.471161324	0.113207753	1.33620246	31.90942187
0.251364241	0.000342357	0.481606482	0.134331251	1.382563429	36.75725569
0.255779668	0.000406189	0.492272864	0.158178066	1.43052202	42.22364304
0.260054199	0.000484011	0.503960447	0.188039801	1.480118783	48.19948522
0.264106143	0.000575553	0.515915459	0.221952272	1.531458469	55.32891344
0.267989246	0.000680989	0.528955793	0.260459593	1.584554738	63.1595228
0.272086864	0.000805649	0.542340374	0.308677315	1.639504393	72.29979404
0.27692797	0.00098348	0.555196682	0.362156909	1.696330469	82.24497192
0.282336104	0.001208787	0.569238855	0.427441121	1.755152966	94.08160037
0.288227295	0.001480565	0.584538492	0.506721136	1.816004812	107.3968132
0.291867619	0.001751058	0.601181189	0.604773773	1.878923382	121.5747902
0.2973398	0.002027483	0.617335932	0.715494788	1.943999613	137.0497918
0.301668976	0.002381943	0.632928154	0.834011449	2.011364297	155.4673354
0.306470667	0.002822226	0.650938025	0.989516729	2.081047457	175.8686413
0.312844283	0.00337236	0.670479558	1.169091689	2.15315299	199.2248106
0.317282218	0.003991939	0.689525699	1.363570036	2.227752632	225.5254491
0.32288761	0.004676868	0.709131074	1.60538198	2.304928107	254.9423175
0.329637136	0.005704017	0.730422833	1.899796888	2.384758947	287.3933937
0.335458589	0.006666607	0.753483718	2.221949111	2.467368844	324.6536377
0.340568829	0.007852505	0.77728247	2.610718549	2.552791689	364.1958244
0.347050367	0.009288419	0.801839639	3.076937504	2.641227411	411.7004438
0.354190239	0.010902182	0.828457297	3.651792562	2.732700711	463.7813776
0.36121433	0.012982164	0.857271858	4.334650638	2.827244866	515.9331463
0.367241968	0.015333173	0.887064915	5.095206327	2.925115824	577.9660894
0.374820098	0.018402694	0.917895132	5.993387125	3.02635746	646.10432
0.382540073	0.021782214	0.949789612	7.030259636	3.131121114	723.7883237
0.389819023	0.025732402	0.982794218	8.252267143	3.24150532	811.468829
0.397222578	0.030012754	1.016947655	9.693443274	3.351641006	907.0342108
0.404778074	0.035364995	1.052273913	11.3308556	3.456913887	1001.457869
0.413114044	0.041821306	1.088839786	13.30039746		

TABLE A.4: Vapour pressure of ^{3}He

A.3 PHASE DIAGRAM DATA

A.3.1 ^{4}He phase diagram

T(K)	P (bar)	T(K)	P(bar)	T(K)	P(bar)
1.69	0.010887343	1.85	0.02047795	2.01	0.033361774
1.70	0.011104638	1.86	0.021190074	2.02	0.034271172
1.71	0.011406669	1.87	0.021909891	2.03	0.035196155
1.72	0.011843971	1.88	0.022634419	2.04	0.036142938
1.73	0.012467079	1.89	0.023360676	2.05	0.037117737
1.74	0.013326527	1.90	0.024085682	2.06	0.038126768
1.75	0.014435678	1.91	0.024806455	2.07	0.039176248
1.76	0.014906906	1.92	0.025520013	2.08	0.040165788
1.77	0.01541564	1.93	0.026223377	2.09	0.0411749
1.78	0.015958897	1.94	0.026913563	2.10	0.042220036
1.79	0.016533698	1.95	0.027884558	2.11	0.043300343
1.80	0.01713706	1.96	0.028830986	2.12	0.044414971
1.81	0.017766003	1.97	0.029755703	2.13	0.045563069
1.82	0.018417544	1.98	0.030664924	2.14	0.046743786
1.83	0.019088704	1.99	0.031564865	2.15	0.04795627
1.84	0.019776499	2.00	0.032461743		

TABLE A.5: ^{4}He vapour - He-II boundary

T(K)	P (bar)	T(K)	P(bar)
2.16	0.04919967	3.68	0.545007392
2.24	0.06008869	3.76	0.59347801
2.32	0.072295518	3.84	0.645384537
2.40	0.085932421	3.92	0.701330844
2.48	0.101247328	4.00	0.760488327
2.56	0.118163256	4.08	0.82346902
2.64	0.136730686	4.16	0.892580749
2.72	0.156901732	4.24	0.966534151
2.80	0.178132126	4.32	1.044589523
2.88	0.200447752	4.40	1.127590909
2.96	0.224950718	4.48	1.214428155
3.04	0.251586016	4.56	1.307401586
3.12	0.280043979	4.64	1.409177058
3.20	0.310574464	4.72	1.51824008
3.28	0.342807143	4.80	1.631815978
3.36	0.377542182	4.88	1.751904665
3.44	0.415535051	4.96	1.882284872
3.52	0.456253326	5.04	2.022615608
3.60	0.499244117	5.12	2.204946655

TABLE A.6: ^{4}He vapour - He-I boundary

T(K)	P (bar)	T(K)	P(bar)
1.775	28.98998749	2.006	10.75938441
1.786	28.85253148	2.017	10.00180903
1.797	28.55071676	2.028	9.362410123
1.808	28.1183558	2.039	8.695001578
1.819	27.4542973	2.05	7.784049981
1.83	26.68776053	2.061	7.024763861
1.841	25.15673067	2.072	6.527869352
1.852	23.42096743	2.083	6.003672299
1.863	21.89746889	2.094	5.514656084
1.874	21.13164507	2.105	4.961240964
1.885	20.30825516	2.116	4.385812845
1.896	19.48460026	2.127	3.81837488
1.907	18.62998084	2.138	3.22709649
1.918	17.55737411	2.149	2.695280621
1.929	16.733	2.16	2.15338667
1.94	16.01799141	2.171	1.560761916
1.951	15.16362672	2.182	0.960841884
1.962	14.24378931	2.193	0.615178566
1.973	13.35268946	2.204	0.359519063
1.984	12.54082793	2.215	0.112977699
1.995	11.69983996	2.23	0.0553104

TABLE A.7: ^{4}He He-I - He-II boundary

T(K)	P (bar)	T(K)	P(bar)
0.11	24.38809166	0.95	24.69442603
0.17	24.38806567	1.01	24.79190242
0.23	24.38816298	1.07	24.90298199
0.29	24.38820505	1.13	25.02840977
0.35	24.38803498	1.19	25.19213929
0.41	24.38769975	1.25	25.4245368
0.47	24.3878713	1.31	25.68207461
0.53	24.38969722	1.37	25.85871129
0.59	24.39825679	1.43	25.88381565
0.65	24.42072678	1.49	26.40458151
0.71	24.45946834	1.55	26.93627771
0.77	24.50668539	1.61	27.58470614
0.83	24.55717091	1.67	28.26326066
0.89	24.61601228		

TABLE A.8: ^{4}He solid - He-II boundary

T(K)	P (bar)
1.73	28.88533512
1.79	29.90570206
1.85	31.19100348
1.91	32.74984185
1.97	34.70493343
2.03	36.80575332
2.09	38.95830369
2.15	41.15278645
2.21	43.36645261
2.27	45.66216025
2.33	48.05068672
2.39	50.51496903
2.45	53.04032615
2.51	55.67141378
2.57	58.41287094
2.63	61.1865465
2.69	64.00597133
2.75	66.81692014
2.81	69.65249806
2.87	72.43405458
2.93	75.17400384
2.99	77.87143141
3.05	80.44990252
3.11	82.93605203
3.17	85.62571217
3.23	89.05610597
3.29	91.98540183

TABLE A.9: ^{4}He solid - He-I boundary

A.3.2 ^{3}He phase diagram

Liquid - Vapour transition data can be taken from Table A.4

T(K)	P (bar)
0.0001	33.75309007
0.0002584	33.75535194
0.0004168	33.75760786
0.0005752	33.75985188
0.0007336	33.76207807
0.000892	33.76428047
0.0010504	33.76645314
0.0012088	33.76859014
0.0013672	33.7706855
0.0015256	33.7727333
0.001684	33.77472758
0.0018424	33.7766624
0.0020008	33.77853181
0.0021592	33.78032986
0.0023176	33.78205061
0.002476	33.78368812
0.0026344	33.78523642

TABLE A.10: ^{3}He solid - superfluid boundary. Until 0.00185 K transition between solid and superfluid phase B, then transition with superfluid phase A.

T(K)	P (bar)	T(K)	P (bar)	T(K)	P (bar)
0.0027136	33.78597527	0.0031096	33.78928673	0.0035056	33.79190214
0.0225334	33.48993909	0.0229294	33.47652309	0.0233254	33.46309095
0.0423532	32.98319685	0.0427492	32.97463254	0.0431452	32.96566087
0.062173	32.39382844	0.062569	32.38087678	0.062965	32.36674722
0.0819928	31.78285165	0.0823888	31.77601808	0.0827848	31.76869656
0.1018126	31.36317615	0.1022086	31.36247621	0.1026046	31.3617584
0.1216324	31.04570913	0.1220284	31.03755351	0.1224244	31.02795501
0.1414522	30.75529021	0.1418482	30.74942948	0.1422442	30.74274045
0.161272	30.53178442	0.161668	30.52632639	0.162064	30.52108087
0.1810918	30.34384509	0.1814878	30.33628446	0.1818838	30.3278859
0.2009116	30.17661884	0.2013076	30.17213838	0.2017036	30.16857935
0.2207314	30.03913687	0.2211274	30.02851544	0.2215234	30.0192216
0.2405512	29.96687439	0.2409472	29.96139512	0.2413432	29.95567558
0.260371	29.8687677	0.260767	29.86658699	0.261163	29.86570829
0.2801908	29.80717502	0.2805868	29.80216521	0.2809828	29.79797609
0.3000106	29.77361247	0.3004066	29.77113988	0.3008026	29.76846524
0.3198304	29.74431291	0.3202264	29.7448336	0.3206224	29.74490461
0.3396502	29.74067793	0.3400462	29.74275808	0.3404422	29.744227
0.35947	29.78323188	0.359866	29.78245704	0.360262	29.7813406
0.3792898	29.85370789	0.3796858	29.85316236	0.3800818	29.85167333
				0.3999016	29.96132499

TABLE A.11: ^{3}He solid - liquid boundary

T(K)	P (bar)	T(K)	P (bar)
0.00188101	33.37043864	0.00191071	32.8846005
0.00189685	33.09091291	0.00192655	32.62360247
0.00191269	32.8542442	0.00194239	32.2136779
0.00192853	32.58250388	0.00195823	31.77986828
0.00194437	32.15547404	0.00197407	31.38405462
0.00196021	31.73096033	0.00198991	30.90803938
0.00197605	31.3308221	0.00200575	30.3122018
0.00199189	30.83883443	0.00202159	29.73682522
0.00200773	30.23424173	0.00203743	29.25846972
0.00202357	29.67137904	0.00205327	28.71679079
0.00203941	29.20139288	0.00206911	28.00756737
0.00205525	28.63306397	0.00208495	27.357126
0.00207109	27.91618056	0.00210079	26.89547305
0.00208693	27.29066046	0.00211663	26.31445131
0.00210277	26.83827952	0.00213247	25.5295719
		0.00214831	24.87225106

TABLE A.12: ^{3}He superfluid phase B - superfluid phase A boundary

T(K)	P (bar)	T(K)	P (bar)
0.00228097	20.9897948	0.00228394	21.000
0.002291068	21.04291716	0.002294038	21.05444937
0.002301166	21.08373157	0.002304136	21.09542816
0.002311264	21.11857852	0.002314234	21.12500085
0.002321362	21.1295745	0.002324332	21.1293676
0.00233146	21.15404684	0.00233443	21.16849982
0.002341558	21.21348698	0.002344528	21.23503901
0.002351656	21.28880585	0.002354626	21.31057524
0.002361754	21.39750401	0.002364724	21.45636904
0.002371852	21.69009594	0.002374822	21.77395185
0.00238195	22.00524947	0.00238492	22.10746036
0.002392048	22.3456381	0.002395018	22.43522745
0.002402146	22.68314612	0.002405116	22.75201293
0.002412244	22.80429711	0.002415214	22.80239967
0.002422342	22.8187586	0.002425312	22.85206908
		0.00243541	23.1988383

TABLE A.13: ^{3}He superfluid phase A - liquid boundary

T(K)	P (bar)	T(K)	P (bar)
0.00090298	0.096781958	0.00092674	0.292205926
0.00096436	0.81472645	0.00098812	1.33196852
0.00102574	2.093614144	0.0010495	2.45474201
0.00108712	3.091013565	0.00111088	3.478099286
0.0011485	3.955934791	0.00117226	4.2577749
0.00120988	4.657288119	0.00123364	4.946814621
0.00127126	5.360355948	0.00129502	5.682445874
0.00133264	6.228546638	0.0013564	6.600994175
0.00139402	7.177007784	0.00141778	7.417069953
0.0014554	7.905221982	0.00147916	8.234452319
0.00151678	8.643046012	0.00154054	8.875479904
0.00157816	9.236528762	0.00160192	9.493580467

TABLE A.14: ^{3}He superfluid phase B - liquid boundary

APPENDIX B

Python Codes

B.1 FILM BREAKER CODE

This code generates the plots in Section 4.3.1. The user may enter an orifice diameter and the code will calculate the load curves given in the case of both molecular and continuum flow. As was discussed in Section 4.3.1, the actual physical load curve will tend towards the continuum model at higher pressures. As such, the authors recommend using the load curve from the continuum model as a conservative guide in designing the film breaker.

```
1  ###################
2
3  # Import packages
4
5  import numpy as np
6  import scipy as sp
7  import matplotlib
8  from matplotlib import pyplot as plt
9
10 ###################
11
12 # Helium properties
13
14 # Helium data from Donnelly, Russell J., and Carlo F. Barenghi.
15 # "The observed properties of liquid helium at the saturated vapor pressure
       ."
16 # Journal of Physical and Chemical Reference Data 27.6 (1998): 1217-1274.
17
18 # Vapour pressure (T in K, P in Pa)
19
20 T_DONNELLY_P = np.array([0.65, 0.7, 0.75, 0.8, 0.85, 0.9, 0.95, 1, 1.05,
       1.1, 1.15, 1.2, 1.25, 1.3, 1.35, 1.4, 1.45, 1.5, 1.55, 1.6, 1.65,
       1.7, 1.75, 1.8, 1.85, 1.9, 1.95, 2, 2.05, 2.1, 2.15, 2.2, 2.25, 2.3,
       2.35, 2.4, 2.45, 2.5, 2.55, 2.6, 2.65, 2.7, 2.75, 2.8, 2.85, 2.9,
       2.95, 3, 3.05, 3.1, 3.15, 3.2, 3.25, 3.3, 3.35, 3.4, 3.45, 3.5, 3.55,
        3.6, 3.65, 3.7, 3.75, 3.8, 3.85, 3.9, 3.95, 4, 4.05, 4.1, 4.15, 4.2,
        4.25, 4.3, 4.35, 4.4, 4.45, 4.5, 4.55, 4.6, 4.65, 4.7, 4.75, 4.8,
       4.85, 4.9, 4.95, 5, 5.05, 5.1])
21
22 P_DONNELLY_P = np.array([1.10E-01, 2.92E-01, 6.89E-01, 1.48E+00, 2.91E+00,
       5.38E+00, 9.38E+00, 1.56E+01, 2.48E+01, 3.80E+01, 5.65E+01, 8.15E+01,
        1.15E+02, 1.58E+02, 2.13E+02, 2.82E+02, 3.67E+02, 4.72E+02, 5.97E
       +02, 7.47E+02, 9.23E+02, 1.13E+03, 1.37E+03, 1.64E+03, 1.95E+03, 2.30
       E+03, 2.69E+03, 3.13E+03, 3.61E+03, 4.14E+03, 4.72E+03, 5.34E+03,
       6.01E+03, 6.73E+03, 7.51E+03, 8.35E+03, 9.26E+03, 1.02E+04, 1.13E+04,
```

```
        1.24E+04, 1.36E+04, 1.48E+04, 1.61E+04, 1.76E+04, 1.91E+04, 2.06E
        +04, 2.23E+04, 2.41E+04, 2.59E+04, 2.78E+04, 2.99E+04, 3.20E+04, 3.43
        E+04, 3.66E+04, 3.90E+04, 4.16E+04, 4.43E+04, 4.71E+04, 4.99E+04,
        5.30E+04, 5.61E+04, 5.94E+04, 6.27E+04, 6.63E+04, 6.99E+04, 7.37E+04,
        7.76E+04, 8.16E+04, 8.58E+04, 9.01E+04, 9.46E+04, 9.92E+04, 1.04E
        +05, 1.09E+05, 1.14E+05, 1.19E+05, 1.25E+05, 1.30E+05, 1.36E+05, 1.42
        E+05, 1.48E+05, 1.54E+05, 1.61E+05, 1.67E+05, 1.74E+05, 1.81E+05,
        1.89E+05, 1.96E+05, 2.04E+05, 2.12E+05])
23
24 # Latent heat (T in K, L in J/mol)
25
26 T_DONNELLY_L = np.array([0, 0.05, 0.1, 0.15, 0.2, 0.25, 0.3, 0.35, 0.4,
        0.45, 0.5, 0.55, 0.6, 0.65, 0.7, 0.75, 0.8, 0.85, 0.9, 0.95, 1, 1.05,
        1.1, 1.15, 1.2, 1.25, 1.3, 1.35, 1.4, 1.45, 1.5, 1.55, 1.6, 1.65,
        1.7, 1.75, 1.8, 1.85, 1.9, 1.95, 2, 2.05, 2.1, 2.18, 2.2, 2.25, 2.3,
        2.35, 2.4, 2.45, 2.5, 2.55, 2.6, 2.65, 2.7, 2.75, 2.8, 2.85, 2.9,
        2.95, 3, 3.05, 3.1, 3.15, 3.2, 3.25, 3.3, 3.35, 3.4, 3.45, 3.5, 3.55,
        3.6, 3.65, 3.7, 3.75, 3.8, 3.85, 3.9, 3.95, 4, 4.05, 4.1, 4.15, 4.2,
        4.25, 4.3, 4.35, 4.4, 4.45, 4.5, 4.55, 4.6, 4.65, 4.7, 4.75, 4.8,
        4.85, 4.9, 4.95, 5, 5.05, 5.1, 5.15])
27
28 L_DONNELLY_L = np.array([59.83, 60.87, 61.91, 62.95, 64, 65.04, 66.08,
        67.13, 68.17, 69.21, 70.24, 71.28, 72.31, 73.33, 74.35, 75.37, 76.38,
        77.38, 78.37, 79.35, 80.33, 81.3, 82.26, 83.21, 84.14, 85.06, 85.97,
        86.87, 87.73, 88.56, 89.35, 90.09, 90.77, 91.38, 91.91, 92.36,
        92.72, 92.98, 93.13, 93.16, 93.07, 92.8, 92.27, 90.75, 90.87, 91.15,
        91.43, 91.71, 91.98, 92.24, 92.5, 92.74, 92.97, 93.18, 93.38, 93.56,
        93.71, 93.85, 93.96, 94.05, 94.11, 94.14, 94.14, 94.11, 94.05, 93.94,
        93.8, 93.63, 93.41, 93.14, 92.84, 92.48, 92.08, 91.63, 91.13, 90.58,
        89.97, 89.31, 88.59, 87.82, 87, 86.13, 85.2, 84.22, 83.19, 82.11,
        80.98, 79.8, 78.57, 77.27, 75.86, 74.32, 72.59, 70.64, 68.44, 65.94,
        63.11, 59.9, 56.28, 52.22, 47.67, 42.59, 36.95, 29.34])
29
30 # Define temperature range for model
31
32 T_model = np.arange(0.65, 1., 0.0001)
33
34 # Define latent heat and vapour pressure as function of temperature
35
36 L_model_fT = np.interp(T_model, T_DONNELLY_L, L_DONNELLY_L)
37 P_model_fT = np.interp(T_model, T_DONNELLY_P, P_DONNELLY_P)
38
39 # Additional physical properties
40
41 M = 4.E-3 # Molar mass
42 k = 5./3. # Ratio of specific heat capacities
43 R = 2077. # Specific gas constant
44
45 ###################
46
47 # Geometrical properties of system
48
49 r = 500.E-6 / 2. # Orifice radius in m
50
51 ###################
52
53 # Models
54
55 # Model A: Continuum flow
56
57 # Mass flow, kg/s
58
59 mA = np.pi * (r**2) * ((P_model_fT)/(T_model * R)) * ((2/(k-1))**0.5) *
        ((2/(k+1))**(1/(k-1))) * ((1-(2/(k+1)))**0.5) * (((k*T_model*R))
        **0.5)
60
61 # Heat lift, W
```

```
62
63 QA = ( mA * L_model_fT ) / M
64
65 # Model B: Molecular flow
66
67 # Mass flow, kg/s
68
69 mB = np.pi * (r**2) * ((P_model_fT)/(T_model * R)) * (((k*T_model*R)/M)
       **0.5) * ((1/(2*np.pi*k))**0.5)
70
71 # Heat lift, W
72
73 QB = ( mB * L_model_fT ) / M
74
75 ##################
76
77 # Plotting
78
79 # Plot model A
80
81 plt.plot(QA, T_model, label='Continuum')
82
83 # Plot model B
84
85 plt.plot(QB, T_model, label='Molecular')
86
87 # Plot labels
88
89 plt.xlabel('Applied heat load [W]')
90 plt.ylabel('Pot temperature [K]')
91 plt.title('Load curve bounds in presence of restriction-type film breaker')
92 plt.legend(loc=4)
93
94 # Define data range to plot
95
96 x1,x2,y1,y2 = plt.axis()
97 plt.axis((x1,0.004,y1,y2))
98
99 # Print
100
101 plt.show()
102
103 # End
```

B.2 ^{3}HE FRIDGE DESIGN CODE

This code generates the plots in Figs. 4.20 and 4.22. In order to run this code, it is necessary to create a text file with the vapour pressure curve data for ^{3}He. This can be found in Appendix A. It is possible to change values from line 72 to 91 to obtain a simulation for a different fridge.

```
1 import numpy as np
2 import numpy.ma as ma
3 import matplotlib.pyplot as plt
4 import random as rd
5
6 from scipy.integrate import quad
7 from scipy import interpolate
8
9 #Latent Heat from 0.2K and 1.8K
10 def LatentHeat(x):
11     return (19.63+22.35*x-4.406*x**2)*1e6 #conversion in uJ
12
```

```
13 #Heat Capacity from 0.2K and 1.8K
14 def HeatCapacity(x):
15     return (2.747+0.424*x+1.083*x**2)*1e6 #conversion in uJ
16
17 #Generic formula for thermal conductivity
18 def thermalconductivity(x, a, b):
19     return a*x**b
20
21 def VapourPressure(x):
22     T, P = np.loadtxt('Insert Vapour Pressure Path', unpack = True)
23     f = interpolate.interp1d(T,P)
24     Pvap = f(x)
25     return Pvap
26
27 #Formula for self-cooling loss (from Cheng, Mayer and Page 1996)
28 def exponential(x):
29     f = (2.747+0.424*x+1.083*x**2)/(19.63+22.35*x-4.406*x**2)
30     return f
31
32 def CondensationEff(n, T_ev, T_p, r_p, l_p, r_e, l_e, l_cp, r_cp, r_ec,
       l_ec, Nt):
33     V_p = np.pi*r_p**2*l_p
34     V_ev = np.pi*r_e**2*l_e
35     n_tec = VapourPressure(T_ev)/R*np.pi*r_ec**2*l_ec/T_ev
36     fact_tcp = (T_p-T_ev)/l_cp
37     n_tcp = VapourPressure(T_ev)/R*np.pi*r_cp**2/fact_tcp*(np.log(fact_tcp*
           T_p+T_ev)-np.log(fact_tcp*T_ev+T_ev))
38     fact1 = VapourPressure(T_ev)/R*(V_p/T_p+V_ev/T_ev)
39     return (n-Nt*n_tec-n_tcp-fact1)/(1-VapourPressure(T_ev)*V_m/R/T_p)
40
41 def CondensationEffInv(n, T_ev, T_p, r_p, l_p, r_e, l_e, l_cp, r_cp, r_ec,
       l_ec, Nt):
42     V_p = np.pi*r_p**2*l_p
43     V_ev = np.pi*r_e**2*l_e
44     n_tec = VapourPressure(T_ev)/R*np.pi*r_ec**2*l_ec/T_ev
45     fact_tcp = (T_p-T_ev)/l_cp
46     n_tcp = VapourPressure(T_ev)/R*np.pi*r_cp**2/fact_tcp*(np.log(fact_tcp*
           T_p+T_ev)-np.log(fact_tcp*T_ev+T_ev))
47     fact1 = VapourPressure(T_ev)/R*(V_p/T_p+V_ev/T_ev)
48     return n*(1-VapourPressure(T_ev)*V_m/R/T_p)+Nt*n_tec+n_tcp+fact1
49
50
51 #Boltzmann Constant
52 k_b = 1.38*10**(-14) #in (g*mm^2)/K/s^2
53
54 #Molecular Mass 3He
55 MM = 3.02*1.66*10**(-24) #in g
56
57 #Molar Mass 3He
58 mm = 3.016 #in g/mol
59
60 #Gas Constant
61 R = 8.3144598*1e12 #mm3*Pa/K/mol
62
63 #Molar Volume 3He
64 V_m = 36.84*1e3 #mm3/mol
65
66 #Cycle varying heat load and hold time for different moles quantity
67 n = np.arange(0.3, 1.5, 0.1)
68 T_0 = 1.2 #Condensation Temperature
69 T_f = 0.3 #Operational Temperature
70 T_p = 40. #Cryopump Temperature
71
72
73 #All length in mm
74 l_ec = 150. #Length tube to evaporator
75 r_ec = 3.5 #Radius tube to evaporator
```

```
76 l_cp = 90. #Length tube to pump
77 r_cp = 8.5 #Radius tube to pump
78 r_e = 30. #Radius evaporator
79 h_e = 40. #Height evaporator
80 r_p = 30. #Radius cryopump
81 h_p = 55. #Height cryopump
82 th = 0.5 #Thickness of the tubes
83
84 #Number of tubes between condenser and evaporator
85 Nt = 2.
86
87 Q = np.arange(2., 50., 0.1)
88 t = np.zeros((len(n), len(Q)))
89
90
91 for i in range(len(n)):
92
93     #Condensation efficiency Calculation
94     n_cond = CondensationEff(n[i], T_0, T_p, r_p, h_p, r_e, h_e, l_cp, r_cp
           , r_ec, l_ec, Nt)
95     #Moles after Cooldown
96     I = quad(exponential, T_0, T_f)
97     mol_frac = np.exp(I[0])
98     mol_cool = mol_frac*n_cond
99
100     print mol_cool
101
102     #Computing Pumping Factors
103     x1 = l_ec/r_ec
104     kf1 = (0.98441+0.00466*x1)/(1+0.46034*x1)
105     x2 = l_cp/r_cp
106     kf2 = (0.98441+0.00466*x2)/(1+0.46034*x2)
107
108     for j in range(len(Q)):
109
110         #Thermal conductivity of Stainless Steel from the 3He bath to the
                condenser (value from Barucci 2008)
111         k_Tbath = quad(thermalconductivity, T_f, T_0, args=(55.6, 1.15)) #
                in uW/mm
112
113         sm = 1/(Nt*kf1*r_ec**2)+1/(kf2*r_cp**2)
114         Q_tot = Q[j] + (3.14*Nt*(((r_ec+th)**2-r_ec**2)/l_ec))*k_Tbath[0]
115
116         #mass flow of 3He #3 molar mass of 3He
117         m_dot = Q_tot*mm/LatentHeat(T_f) #in g/s
118         B = 2*m_dot*np.sqrt(k_b/(2*3.14*MM))
119
120         X1 = VapourPressure(T_f)/(np.sqrt(T_f)*B) #Conversion in g/mm/s2
121         if sm<= X1:
122
123             t[i,j] = mol_cool*LatentHeat(T_f)/Q_tot/24/3600.
124
125         else:
126             t[i,j] = np.nan
127
128
129 #Data for Plot
130 z = ma.masked_invalid(t)
131
132 plt.figure(0)
133
134 levels = np.array([5,10, 15, 20, 25, 30, 35])
135
136 CS = plt.contour(Q, n, z, colors = 'k', levels = levels)
137
138 plt.xlabel('$\dot{Q}$ ($\mu$W)')
139 plt.ylabel('n (mol)')
```

```
140 plt.title('Hold Time in days as function of charging moles and load applied
        ')
141 plt.clabel(CS, inline=1, fontsize=10, fmt = '%2.0f')
142
143
144 #Compute the length and radius from the evaporator as a function of Heat
        Load
145 #keeping the tube to the pump constant
146 HeatValues = 5
147 CycValues = 100
148 Q_fix =  np.linspace(10, 50, HeatValues)#uW
149
150 #Pump Tube Dimensions
151 l_cp = 70. #Length tube to pump
152 r_cp = 8.5 #Radius tube to pump
153
154 sm = 10.
155 X1 = 0.
156
157 lect = np.linspace(80, 120, 41)
158 rect = np.linspace(2, 5, 41)
159
160 Q_fix = np.arange(10,50, 10.)
161 Q_tot = np.zeros((41,41))
162
163 for y in range(len(Q_fix)):
164     for i in range(len(lect)):
165
166         print i
167
168         for j in range(len(rect)):
169
170             x1 = lect[j]/rect[i]
171             kf1 = (0.98441+0.00466*x1)/(1+0.46034*x1)
172
173             x2 = l_cp/r_cp
174
175             kf2 = (0.98441+0.00466*x2)/(1+0.46034*x2)
176
177             #Thermal conductivity of Stainless Steel from the 3He bath to
                    the condenser (value from Barucci 2008)
178             k_Tbath = quad(thermalconductivity, T_f, T_0, args=(55.6, 1.15)
                    ) #in uW/mm
179
180             sm = 1/(Nt*kf1*rect[i]**2)+1/(kf2*r_cp**2)
181             Q_tot[i][j] = Q_fix[y] + (3.14*Nt*(((rect[i]+th)**2-rect[i]**2)
                    /lect[j]))*k_Tbath[0]
182
183             #mass flow of 3He #3 molar mass of 3He
184             m_dot = Q_tot[i][j]*mm/LatentHeat(T_f)  #in g/s
185             B = 2*m_dot*np.sqrt(k_b/(2*3.14*MM))
186
187             X1 = VapourPressure(T_f)/(np.sqrt(T_f)*B) #Conversion in g/mm/
                    s2
188             if sm<= X1:
189
190                 Q_tot[i,j] = Q_tot[i][j]
191
192             else:
193                 Q_tot[i][j] = np.nan
194
195     z = ma.masked_invalid(Q_tot)
196
197     plt.figure(1+y)
198     CS = plt.contour(lect, rect, Q_tot, colors = 'k')
199     plt.clabel(CS, inline=1, fontsize=10)
200
```

```
201     plt.xlabel('Length (mm)')
202     plt.ylabel('Radius (mm)')
203     plt.title('Total heat load with an external load of '+str(Q_fix[y])+'$\
          mu$W')
204
205 plt.show()
```

B.3 CRYOSTAT COOLDOWN CODE

The following code uses a finite element approach to model the cooldown of a medium-sized pulse-tube cooled cryostat. Commenting throughout describes the code in detail.

```
1  ###################
2
3  # Cryostat cooldown thermal simulation - Andy May, Oct 2017
4  # andrew.may-3@postgrad.manchester.ac.uk
5
6  ###################
7
8  # The following code models the cooldown of a medium-sized 2 stage cryostat
       by a PT415 pulse tube cryostat
9
10 # The capacity map of the mechanical coolers are used from Green et al.
       (2015)
11 # http://iopscience.iop.org/article/10.1088/1757-899X/101/1/012002/pdf
12
13 # Radiative loads are accounted for using typical shield sizings with
       emissivity consistent with MLI
14
15 # The thermal masses on the 1st and 2nd stages are modeled as lumped
       elements (i.e. isothermal)
16 # These stages are treated as copper with temperature dependent heat
       capacity from Marquardt, Le, and Radebaugh
17 # Cryogenic Material Properties Database, National Institute of Standards
       and Technology Boulder, CO 80303
18
19 ###################
20
21 # Import packages
22
23 import numpy as np
24 import scipy as sp
25 import matplotlib
26 from matplotlib import pyplot as plt
27 from scipy import interpolate
28
29 ###################
30
31 # Model time
32
33 endtime = 5 # hours
34 endtime = endtime * 3600 # secs
35 timestep = 0.1 # secs
36 globaltime = np.array([0.00]) # secs # initialise, will increase by "
       timestep" until reaches "endtime"
37
38 ###################
39
40 # Define physical constants
41
42 stefboltz = 5.67E-8 # W/m2.K4
43
44 ###################
```

```
45
46 # Define initial conditions and parameters for thermal elements
47
48 # Element 0 # Outer vacuum chamber, model as thermal reservoir at 300 K
49
50 T_0 = np.array([300.0]) # K # Initialise temperature array
51
52 # Element 1 # 1st stage, thermal mass
53
54 A_1 = 0.5 # m2 # Surface area
55 eps1 = 0.01 # Surface emissivity
56 m_1 = 12 # kg # Mass
57 T_1 = np.array([300.0]) # K # Initialise temperature array
58
59 # Element 2 # 2nd stage, thermal mass
60
61 A_2 = 0.5 # m2 # Surface area
62 eps2 = 0.01 # Surface emissivity
63 m_2 = 15 # kg # Mass
64 T_2 = np.array([300.0]) # K # Initialise temperature array
65
66 ###################
67
68 # Initialise heat flow arrays
69
70 Qdot1_record = np.array([0.00]) # W # PTC 1st stage heat lift
71 Qdot2_record = np.array([0.00]) # W # PTC 2nd stage heat lift
72
73 QRdot1_record = np.array([0.00]) # W # Radiative load on 1st stage
74 QRdot2_record = np.array([0.00]) # W # Radiative load on 2nd stage
75
76 ###################
77
78 # PTC heat lift data
79
80 T_PTC1 = np.array([31.42, 42.26, 41.72, 41.72, 41.72, 42.26, 49.85, 57.43,
      59.06, 59.06, 60.68, 62.31, 75.31, 85.06, 89.40, 100.23, 115.94,
      122.99, 114.32,141.41, 153.33, 154.95, 163.08, 175.54, 216.18,
      218.89, 236.76, 246.52, 250.85, 286.61, 295.82, 303.95, 308.28]) # K
      # PTC 1st stage temperature
81 T_PTC2 = np.array([2.51, 33.52, 76.26, 144.97, 248.04, 292.46, 2.52, 36.03,
       78.77, 144.97, 224.58, 310.06, 3.35, 40.22, 79.61, 134.08, 201.12,
      282.40, 5.03, 51.12, 97.21, 151.68, 222.07, 315.08, 41.07, 82.96,
      150.00, 232.122, 320.11, 70.39, 130.72, 204.47, 284.92]) # K # PTC 2
      nd stage temperature
82 Q_PTC1 = np.array([0.0, 0.0, 0.0, 0.0, 0.0, 0.0, 50.0, 50.0, 50.0, 50.0,
      50.0, 50.0, 100.0, 100.0, 100.0, 100.0, 100.0, 100.0, 150.0, 150.0,
      150.0, 150.0, 150.0, 150.0, 200.0, 200.0, 200.0, 200.0, 200.0, 250.0,
       250.0, 250.0, 250.0]) # W # PTC 1st stage heat lift
83 Q_PTC2 = np.array([0.0, 25.0, 50.0, 75.0, 100.0, 125.0, 0.0, 25.0, 50.0,
      75.0, 100.0, 125.0, 0.0, 25.0, 50.0, 75.0, 100.0, 125.0, 0.0, 25.0,
      50.0, 75.0, 100.0, 125.0, 0.0, 25.0, 50.0, 75.0, 100.0, 0.0, 25.0,
      50.0, 75.0]) # W # PTC 2nd stage heat lift
84
85 # Interpolated values as f(T_1,T_2)
86
87 Q_PTC1_interp_func = interpolate.interp2d(T_PTC1, T_PTC2, Q_PTC1, kind='
      linear') # W # PTC 1st stage heat lift
88 Q_PTC2_interp_func = interpolate.interp2d(T_PTC1, T_PTC2, Q_PTC2, kind='
      linear') # W # PTC 2nd stage heat lift
89
90 ###################
91
92 # Solver
93
94 while globaltime[-1] < endtime : # Loop until simulation reaches endtime
95
```

```
96      ###################
97
98      # Calculate temperature dependent properties
99
100     # Element 0
101
102     #
103
104     # Element 1
105
106     cp_1 = 10**((-1.91844) + (-0.15973*np.log10(T_1[-1])) + (8.61013*(np.
            log10(T_1[-1])**2)) + (-18.99640*(np.log10(T_1[-1])**3)) +
            (21.96610*(np.log10(T_1[-1])**4)) + (-12.73280*(np.log10(T_1[-1])
            **5) + (3.54322*(np.log10(T_1[-1])**6))) + (-0.37970*(np.log10(
            T_1[-1])**7)))
107     Cp_1 = cp_1 * m_1 # J/K
108
109     # Element 2
110
111     cp_2 = 10**((-1.91844) + (-0.15973*np.log10(T_2[-1])) + (8.61013*(np.
            log10(T_2[-1])**2)) + (-18.99640*(np.log10(T_2[-1])**3)) +
            (21.96610*(np.log10(T_2[-1])**4)) + (-12.73280*(np.log10(T_2[-1])
            **5) + (3.54322*(np.log10(T_2[-1])**6))) + (-0.37970*(np.log10(
            T_2[-1])**7)))
112     Cp_2 = cp_2 * m_2 # J/K
113
114     ###################
115
116     # Calculate heat transfer
117
118     # Element 0
119
120     #
121
122     # Element 1
123
124     # PTC heat lift
125
126     Qdot1 = Q_PTC1_interp_func(T_1[-1], T_2[-1]) # W # 1st stage heat lift
            as function of 1st and 2nd stage temperatures
127     Q1 = Qdot1 * timestep # J # Heat removed in timestep
128     Qdot1_record = np.append(Qdot1_record,Qdot1) # W # Record for plotting
129
130     # Radiative load
131
132     QRdot1 = A_1 * eps1 * stefboltz * ((T_0[-1] **4) - (T_1[-1] **4)) # W #
             Rad load on 1st stage
133     QR1 = QRdot1 * timestep # J # Radiative heat added in timestep
134     QRdot1_record = np.append(QRdot1_record,QRdot1) # W # Record for
            plotting
135
136     # Element 2
137
138     # PTC heat lift
139
140     Qdot2 = Q_PTC2_interp_func(T_1[-1], T_2[-1]) # W # 2nd stage heat lift
            as function of 1st and 2nd stage temperatures
141     Q2 = Qdot2 * timestep # J # Heat removed in timestep
142     Qdot2_record = np.append(Qdot2_record,Qdot2) # W # Record for plotting
143
144     # Radiative load
145
146     QRdot2 = A_2 * eps2 * stefboltz * ((T_1[-1] **4) - (T_2[-1] **4)) # W #
             Rad load on 2nd stage
147     QR2 = QRdot2 * timestep # J # Radiative heat added in timestep
148     QRdot2_record = np.append(QRdot2_record,QRdot2) # W # Record for
            plotting
```

```
149
150     ###################
151
152     # Calculate temperatures at end of timestep
153
154     # Element 0
155
156     T_0_new = T_0[-1] # K # Thermal reservoir, hence temperature unchanged
157     T_0 = np.append(T_0,T_0_new) # Add new temperature to array
158
159
160     # Element 1
161
162     T_1_new = ( T_1[-1] ) + ( ( -Q1 + QR1 -QR2 ) / Cp_1) # K # Temperature
            change from net heat transfer divided by total heat capacity
163     T_1 = np.append(T_1,T_1_new) # K # Add new temperature to array
164
165     # Element 2
166
167     T_2_new = ( T_2[-1] ) + ((-Q2+QR1) / Cp_2) # K # Temperature change
            from net heat transfer divided by total heat capacity
168     T_2 = np.append(T_2,T_2_new) # K # Add new temperature to array
169
170     ###################
171
172     # Timekeeping
173
174     time_new = globaltime[-1] + timestep # Move time on by one timestep
175     globaltime = np.append(globaltime,time_new) # Add new time to array
176
177     ###################
178
179 ###################
180
181 # Plotting
182
183 # Plot temperatures over time
184
185 plt.figure() # Initialise figure
186
187 globaltime = globaltime / 3600 # hours
188
189 plt.plot(globaltime, T_0, label='Vacuum chamber') # Plot vacuum chamber
        temperature
190 plt.plot(globaltime, T_1, label='1st stage') # Plot 1st stage temperature
191 plt.plot(globaltime, T_2, label='2nd stage') # Plot 2nd stage temperature
192
193 plt.xlabel('Time [hours]') # x-axis label
194 plt.ylabel('Temperature [K]') # y-axis label
195 plt.title('Thermal simulation of Manchester cryostat cooldown') # Title
196 plt.legend(loc=1) # Legend position
197
198 x1,x2,y1,y2 = plt.axis() # Define axis ranges
199 plt.axis((x1,x2,y1,310)) # Define axis ranges
200
201 # Plot heat transfer rates
202
203 plt.figure() # Initialise figure
204
205 plt.plot(globaltime, Qdot1_record, label='PTC1 heat lift') # Plot PTC 1st
        stage heat lift
206 plt.plot(globaltime, Qdot2_record, label='PTC2 heat lift') # Plot PTC 2nd
        stage heat lift
207 plt.plot(globaltime, QRdot1_record, label='Radiative load on 1st stage') #
        Plot 1st stage radiative heat load
208 plt.plot(globaltime, QRdot2_record, label='Radiative load on 2nd stage') #
        Plot 2nd stage radiative heat load
```

```
209
210 plt.xlabel('Time [hours]') # x-axis label
211 plt.ylabel('Qdot [W]') # y-axis label
212 plt.title('Thermal simulation of Manchester cryostat cooldown') # Title
213 plt.legend(loc=1) # Legend position
214
215 x1,x2,y1,y2 = plt.axis() # Define axis ranges
216 plt.axis((0,x2,0,y2)) # Define axis ranges
217
218 plt.show() # Display plots
219
220 ###################
221
222 # End
223
224 ###################
```

Bibliography

[1] I.N. Adamenko and I.M. Fuks. Roughness and thermal resistance of the boundary between a solid and liquid helium. *Sov. Phys. JETP*, 32(6):1123–1129, 1971.

[2] J.F. Allen and A.D. Misener. The properties of flow of liquid He II. *Proceedings of the Royal Society of London. Series A, Mathematical and Physical Sciences*, pages 467–491, 1939.

[3] A.C. Anderson, D.O. Edwards, W.R. Roach, R.E. Sarwinski, and J.C. Wheatley. Thermal and magnetic properties of dilute solutions of ^{3}He in ^{4}He at low temperatures. *Phys. Rev. Lett.*, 17:367, 1966.

[4] K.R. Atkins. The flow of liquid helium II through wide capillaries. *Proceedings of the Physical Society. Section A*, 64(9):833, 1951.

[5] J. Aumont, S. Banfi, P. Battaglia, E.S. Battistelli, A. Bau, B. Belier, D. Bennett, L. Berge, J.-Ph. Bernard, M. Bersanelli, et al. QUBIC technological design report. *arXiv preprint 1609.04372*, 2016.

[6] J. Bardeen, L.N. Cooper, and J.R. Schrieffer. Theory of superconductivity. *Phys. Rev.*, 108(5):1175, 1957.

[7] J.D. Beckenstein. Generalized second law of thermodynamics in black-hole physics. *Phys. Rev. D.*, 9(12):3292, 1974.

[8] A. Benoit and S. Pujol. A dilution refrigerator insensitive to gravity. *Physica B*, 169(1-4):457–458, 1991.

[9] R.S. Bhatia, S.T. Chase, S.F. Edgington, J. Glenn, W.C. Jones, A.E. Lange, B. Maffei, A.K. Mainzer, P.D. Mauskopf, B.J. Philhour, et al. A three-stage helium sorption refrigerator for cooling of infrared detectors to 280 mk. *Cryogenics*, 40(11):685–691, 2000.

[10] S. Brunauer. The Adsorption of Gases and Vapors. Physical Adsorption. *Oxford University Press*, 1:528, 1943.

[11] S. Brunauer, P. Emmett, and E. Teller. Adsorption of gases in multimolecular layers. *Journal of the American Chemical Society*, 60(2):309–319, 1938.

[12] G. Chaudhry. *Thermodynamic properties of liquid 3He-4He mixtures between 0.15 K and 1.8 K.* PhD thesis, Massachusetts Institute of Technology, 2009.

[13] G. Chaudhry, A. Volpe, P. Camus, S. Triqueneaux, and G. Vermeulen. A closed-cycle dilution refrigerator for space applications. *Cryogenics*, 52:471–477, 2012.

[14] E.S. Cheng, S.S. Meyer, and L.A. Page. A high capacity 0.23 K 3He refrigerator for balloon-borne payloads. *Review of scientific instruments*, 67(11):4008–4016, 1996.

[15] R.D. Cook, D.S. Malkus, M.E. Plesha, and R.J. Witt. *Concepts and applications of finite element analysis*, volume 4. Wiley New York, 1974.

[16] G. Coppi, P. de Bernardis, A.J. May, S. Masi, M. McCulloch, S.J. Melhuish, and L. Piccirillo. Developing a long duration 3He fridge for the LSPE-SWIPE instrument. In *SPIE Astronomical Telescopes+Instrumentation*, pages 991265–1–991265–8. International Society for Optics and Photonics, 2016.

[17] G. Dall'Oglio, L. Martinis, G. Morgante, and L. Pizzo. An improved 3He refrigerator. *Cryogenics*, 37(1):63–64, 1997.

[18] G. Dall'Oglio, L. Pizzo, L. Piccirillo, and L. Martinis. New 3He/4He refrigerator. *Cryogenics*, 31(1):61–63, 1991.

[19] J.G. Dash and H.A. Boorse. Transport rates of the Helium II film over various surfaces. *Physical Review*, 82(6):851, 1951.

[20] M.J. Devlin, S.R. Dicker, J. Klein, and M.P. Supanich. A high capacity completely closed-cycle 250 mk 3He refrigeration system based on a pulse tube cooler. *Cryogenics*, 44(9):611–616, 2004.

[21] R.J. Donnelly and C.F. Barenghi. The observed properties of liquid helium at the saturated vapor pressure. *Journal of Physical and Chemical Reference Data*, 27(6):1217, 1998.

[22] L. Duband and B. Collaudin. Sorption coolers development at CEA-SBT. *Cryogenics*, 39(8):659–663, 1999.

[23] S. Dushman. *Scientific Foundation of Vacuum Technology.* John Wiley & Sons, 1962.

[24] V.S. Edel'man. A dilution refrigerator with condensation pump. *Cryogenics*, 12:385–387, 1972.

[25] V.S. Edel'man. Operation of a dilution refrigerator in a micromode. *Instruments and Experimental Techniques*, 45(3):420–425, 2002.

[26] A. Einstein. *Autobiographical Notes. A Centennial Edition.* Open Court Publishing Company, 1979.

[27] B.H. Flowers and E. Mendoza. *Properties of Matter.* Wiley, 1970.

[28] G. Frossati. Experimental techniques: Methods for cooling below 300 mK. *Journal of Low Temperature Physics*, 87(3/4):595–633, 1992.

[29] J. Gleick. The Information: A History, A Theory, A Flood [Book Review]. *IEEE Transactions on Information Theory*, 57(9):6332–6333, 2011.

[30] A. Graziani, G. Dall'Oglio, L. Pizzo, and L. Sabbatini. A new generation of ^{3}He refrigerators. *Cryogenics*, 43(12), 2003.

[31] M.A. Green, C. Wang, and A.F. Zeller. Second stage cooling from a Cryomech PT415 cooler at second stage temperatures up to 300 K with cooling on the first-stage from 0 to 250 W. In *CEC 2015*, pages 1–15, 2015.

[32] D.S. Greywall. Thermal-conductivity measurements in liquid ^{4}He below 0.7 K. *Phys. Rev. B*, 23:2152–2168, 1981.

[33] S.W. Hawking. Black hole explosions. *Nature*, 248(5443):30–31, 1974.

[34] S.W. Hawking. Black holes and thermodynamics. *Phys. Rev. D*, 13(2):191, 1976.

[35] C.V. Heer and J.G. Daunt. A Contribution to the Theory of Bose-Einstein Liquids. *Physical Review*, 81(3):447–454, 1951.

[36] R. Herrmann, A.V. Ofitserov, I.N. Khlyustikov, and V.S. Edel'man. A portable dilution refrigerator. *Instruments and Experimental Techniques*, 48(5):693–702, 2005.

[37] Y.H. Huang and G.B. Chen. A practical vapor pressure equation for helium-3 from 0.01K to the critical point. *Cryogenics*, 46(12):833–839, 2006.

[38] Y.H. Huang, G.B. Chen, and V. Arp. Debye equation of state for fluid helium-3. *The Journal of Chemical Physics*, 125(5):054505, 2006.

[39] Y.H. Huang, G.B. Chen, B.H. Lai, and S.Q. Wang. p-H and T-S diagrams of 3He from 0.2K to 20K. *Cryogenics*, 45(10):687–693, 2005.

[40] Y.H. Huang, G.B. Chen, and R.Z. Wang. Thermodynamic Diagrams of 3He from 0.2 K to 300 K Based Upon its Debye Fluid Equation of State. *International Journal of Thermophysics*, 31(4-5):774–783, 2010.

[41] T. Hughes. *The Finite Element Method: Linear Static and Dynamic Finite Element Analysis.* Courier Corporation, 2012.

[42] J. E. Jones. On the determination of molecular fields. ii. from the equation of state of a gas. *Proceedings of the Royal Society of London A: Mathematical, Physical and Engineering Sciences*, 106(738):463–477, 1924.

[43] G.M. Klemencic, P.A.R. Ade, S. Chase, R. Sudiwala, and A.L. Woodcraft. A continuous dry 300 mK cooler for THz sensing applications. *Review of Scientific Instruments*, 87(4):045107, 2016.

[44] J.G.M. Kuerten, C.A.M. Castelijns, Ph. Camus, A. Benoit, and G. Guyot. Comprehensive theory of flow properties of ^{3}He moving through superfluid ^{4}He in capillaries. *Phys. Rev. Lett.*, 46(21):2288, 1986.

[45] J.G.M. Kuerten, C.A.M. Castelijns, A.T.A.M. de Waele, and H.M. Gijsman. Thermodynamic properties of liquid 3He-4He mixtures at zero pressure for temperatures below 250 mK and 3He concentrations below 8%. *Cryogenics*, 25(8):419–443, 1985.

[46] L. Landau. On the theory of superfluidity. *Physical Review*, 75(5):884, 1949.

[47] J. Lau, M. Benna, M. Devlin, S. Dicker, and L. Page. Experimental tests and modeling of the optimal orifice size for a closed cycle 4 He sorption refrigerator. *Cryogenics*, 46(11):809–814, 2006.

[48] H.M. Ledbetter. *Materials at Low Temperatures.* ASM International, 1983.

[49] D. M. Lee and Henry A. Fairbank. Heat transport in liquid he^3. *Phys. Rev.*, 116:1359–1364, 1959.

[50] H. London. *Proceedings of the International Conference on Low Temperature Physics*, 1951.

[51] H. London, G. R. Clarke, and E. Mendoza. Osmotic Pressure of He 3 in Liquid He 4 , with Proposals for a Refrigerator to Work below 1K. *Physical Review*, 128(5):1992–2005, 1962.

[52] O.V. Lounasmaa. *Experimental Principles and Methods Below 1K.* Academic Press, 1974.

[53] D. Lynden-Bell and R.M. Lynden-Bell. On the negative specific heat paradox. *Monthly Notices of the Royal Astronomical Society*, 181(3):405–419, 1977.

[54] F. Mandl. *Statistical Physics.* John Wiley & Sons, 1988.

[55] D. Martins, L. Ribeiro, D. Lopes, I. Catarino, I.A.A.C. Esteves, J.P.B. Mota, and G. Bonfait. Sorption characterization and actuation of a gas-gap heat switch. *Sensors and Actuators A: Physical*, 171(2):324–331, 2011.

[56] A.J. May, P.G. Calisse, G. Coppi, V. Haynes, L. Martinis, M.A. McCulloch, S.J. Melhuish, and L. Piccirillo. Sorption-cooled continuous miniature dilution refrigeration for astrophysical applications. In *SPIE Astronomical Telescopes+ Instrumentation*, pages 991266–991266. International Society for Optics and Photonics, 2016.

[57] A.J. May, G. Coppi, L. Martinis, S.J. Melhuish, and L. Piccirillo. A closed-cycle convective ^{4}He heat switch. *In Preparation*, 2017.

[58] A.J. May, G. Coppi, and L. Piccirillo. A superfluid film breaker for high power ^{4}He sorption coolers. *In Preparation*, 2017.

[59] A.J. May, G. Coppi, and L. Piccirillo. The QUBIC 1 K refrigerator. *In Preparation*, 2017.

[60] S.J. Melhuish, L. Martinis, and L. Piccirillo. A tiltable single-shot miniature dilution refrigerator for astrophysical applications. *Cryogenics*, 55:63–67, 2013.

[61] S.J. Melhuish, A.J. May, G. Coppi, and L. Piccirillo. A continuous 350 mK sorption cooler using convective heat switches. *In Preparation*, 2017.

[62] S.J. Melhuish, C. Stott, A. Ariciu, L. Martinis, M. McCulloch, L. Piccirillo, D. Collison, F. Tuna, and R. Winpenny. A sub-Kelvin cryogen-free EPR system. *Journal of magnetic resonance*, 282:83–88, 2017.

[63] V.A. Mikheev, V.A. Maidanov, and N.P. Mikhin. Compact dilution refrigerator with a cryogenic circulation cycle of He3. *Cryogenics*, 24:190, 1984.

[64] V. Musilova, P. Hanzelka, T. Kralik, and A. Srnka. Low temperature radiative properties of materials used in cryogenics. *Cryogenics*, 45(8):529–536, 2005.

[65] N.H. Pennings, K.W. Taconis, and R. de Bruyn Ouboter. An improved version of the ^{3}He-^{4}He refrigerator through which ^{4}He is circulated. *Cryogenics*, 14(1):53–54, 1974.

[66] F. Pobell. *Matter and Methods at Low Temperatures*. Springer, 1995.

[67] R. Radebaugh. Thermodynamic properties of HE3-HE4 solutions with applications to the HE3-HE4 dilution refrigerator. *NBS Technical Note*, 392, 1967.

[68] A. Ramiere, S. Volz, and J. Amrit. Thermal resistance at a solid/superfluid helium interface. *Nature Materials*, 15(5):512–516, 2016.

[69] P.R. Roach, K.E. Gray, and B.D. Dunlap. Low cost, compact dilution refrigerator. *Japanese Journal of Applied Physics*, 26(S3-2):1727, 1987.

[70] P.R. Roach and B. Helvensteijn. Development of a compact dilution refrigerator for zero gravity operation. *Advances in Cryogenic Engineering*, 35:1045–1053, 1990.

[71] P.R. Roach and B. Helvensteijn. Progress on a microgravity dilution refrigerator. *Cryogenics*, 39(12):1015–1019, 1999.

[72] M.C. Runyan and W.C. Jones. Thermal conductivity of thermally-isolating polymeric and composite structural support materials between 0.3 and 4 K. *Cryogenics*, 48(9):448–454, 2008.

[73] R.G. Scurlock. *History and Origins of Cryogenics.* Clarendon Press, 1992.

[74] D.W. Sedgley, A.G. Tobin, T.H. Batzer, and W.R. Call. Characterization of charcoals for helium cryopumping in fusion devices. *Journal of Vacuum Science & Technology*, 5(2572), 1987.

[75] P.J. Shirron and M.J. DiPirro. Suppression of superfluid film flow in the XRS helium dewar. In *Advances in cryogenic engineering*, pages 949–956. Springer, 1998.

[76] V.A. Sivokon, V.V. Dotsenko, L.A. Pogorelov, and V.I. Sobolev. Dilution refrigerator with condensation pumping. *Cryogenics*, 32:207–210, 1992.

[77] B. Smith and H.A. Boorse. Helium II film transport. I. the role of substrate. *Physical Review*, 98(2):328, 1955.

[78] B. Smith and H.A. Boorse. Helium II film transport. II. the role of surface finish. *Physical Review*, 99(2):346, 1955.

[79] H.F. Stoeckli. A generalization of the Dubinin–Radushkevich equation for the filling of heterogeneous micropore systems. *Journal of Colloid and Interface Science*, 59(1):184–185, 1977.

[80] E.T. Swartz and R. O. Pohl. Thermal boundary resistance. *Review of Modern Physics*, 61(3):605, 1989.

[81] B.A. Szabo and I. Babuska. *Finite Element Analysis.* John Wiley & Sons, 1991.

[82] K.V. Taconis, N.H. Pennings, P. Das, and R. de Bruyn Ouboter. A 4he-3he refrigerator through which 4he is circulated. *Physica*, 56(1):168–170, 1971.

[83] Y. Takano. Cooling power of the dilution refrigerator with a perfect continuous counterflow heat exchanger. *Rev. Sci. Instrum.*, 65(5):1667–1674, 1994.

[84] G. Teleberg. *Sorption-cooled Miniature Dilution Refrigerators for Astrophysical Applications.* PhD Thesis, Cardiff University, 2006.

[85] G. Teleberg, S.T. Chase, and L. Piccirillo. A cryogen-free miniature dilution refrigerator for low-temperature detector applications. *Journal of Low Temperature Physics*, 151(3):669–674, 2008.

[86] D.R. Tilley and J. Tilley. *Superfluidity and Superconductivity.* CRC Press, 1990.

[87] R.H. Torii and H.J. Maris. Low-temperature heat switch using unforced convection. *Review of scientific instruments*, 57(4):655–657, 1986.

[88] J.P. Torre and G. Chanin. Miniature liquid-3he refrigerator. *Review of scientific instruments*, 56(2):318–320, 1985.

[89] S. Triqueneaux, L. Sentis, A.T.A.M. de Waele, and H.M. Gijsman. Design and performance of the dilution cooler system for the planck mission. *Cryogenics*, 56(4):288–297, 2006.

[90] S.W. van Sciver. *Helium Cryogenics.* Springer, 2012.

[91] G. Ventura and L. Risegari. *The Art of Cryogenics: Low-temperature Experimental Techniques.* Elsevier, 2008.

[92] G.K. Walters and W.M. Fairbanks. Phase separation in he 3he 4 solutions. *Phys. Rev.*, 103(1):262, 1956.

[93] J.C. Wheatley, R.E. Rapp, and R.T. Johnson. Principles and methods of dilution refrigerators. II. *J. Low Temp. Physics*, 4(1):1–39, 1971.

[94] F.M. White. *Fluid Mechanics.* McGraw-Hill, 2003.

[95] J. Wilks and D.S. Betts. *An Introduction to Liquid Helium.* Clarendon Press, 1987.

Index

Index